Privacy in the Age of Innovation

AI Solutions for Information Security

Ranadeep Reddy Palle
Krishna Chaitanya Rao Kathala

Apress®

Privacy in the Age of Innovation: AI Solutions for Information Security

Ranadeep Reddy Palle
Leander, TX, USA

Krishna Chaitanya Rao Kathala
Amherst, MA, USA

ISBN-13 (pbk): 979-8-8688-0460-1
https://doi.org/10.1007/979-8-8688-0461-8

ISBN-13 (electronic): 979-8-8688-0461-8

Managing Director, Apress Media LLC: Welmoed Spahr
Acquisitions Editor: Celestin Suresh John
Development Editor: Laura Berendson
Coordinating Editor: Gryffin Winkler

Cover image designed by Freepik (www.freepik.com)

Distributed to the book trade worldwide by Springer Science+Business Media LLC, 1 New York Plaza, Suite 4600, New York, NY 10004. Phone 1-800-SPRINGER, fax (201) 348-4505, e-mail orders-ny@springer-sbm.com, or visit www.springeronline.com. Apress Media, LLC is a California LLC and the sole member (owner) is Springer Science + Business Media Finance Inc (SSBM Finance Inc). SSBM Finance Inc is a **Delaware** corporation.

For information on translations, please e-mail booktranslations@springernature.com; for reprint, paperback, or audio rights, please e-mail bookpermissions@springernature.com.

Apress titles may be purchased in bulk for academic, corporate, or promotional use. eBook versions and licenses are also available for most titles. For more information, reference our Print and eBook Bulk Sales web page at http://www.apress.com/bulk-sales.

Any source code or other supplementary material referenced by the author in this book is available to readers on GitHub (https://github.com/Apress). For more detailed information, please visit https://www.apress.com/gp/services/source-code.

If disposing of this product, please recycle the paper

Table of Contents

About the Authors

Ranadeep Reddy Palle has over 12 years of experience and specializes in the development of distributed and scalable microservices. His proficiency extends to deploying these microservices on major cloud platforms like AWS and Azure, always prioritizing strict adherence to privacy and cybersecurity standards. Ranadeep has contributed extensively to journals, particularly in critical domains such as machine learning, AI, cloud computing, and data analytics. Holding diverse certifications in both Amazon Web Services (AWS) and Microsoft Azure, he currently serves as a Senior Software Engineer at Zoom in Austin, TX. In this role, he leads a team in implementing cutting-edge solutions and plays a crucial part in the design and development of microservices for consent management, ensuring compliance with global privacy regulations. Ranadeep holds a master's degree in computer science from the University of Houston, Texas.

Krishna Chaitanya Rao Kathala is a prominent researcher in the fields of data science, analytics, machine learning, and AI and a PhD candidate at the University of Massachusetts Amherst. With over eight years of specialized experience in AI, Krishna's expertise covers a wide range of topics within artificial intelligence, including generative AI, ethics, bias, fairness, and safe/secure AI. He is a prolific scholar, having coauthored two books and secured more than five international patents. Additionally, Krishna has authored over eight research papers published in prestigious journals. His influence in the academic and professional realms is widespread, often serving as a judge at more than ten national and international events, including research symposiums and hackathons. Krishna also contributes actively to the academic community through roles on editorial boards, reviews, and program committees at various conferences. Krishna's merit in the field has led to invitations to speak at distinguished international forums such as the United Nations, the Ministry of Foreign Affairs Ukraine, UNESCO, UNDP, United Nations Women, the Office of the United Nations High Commissioner for Human Rights, and the Talent AI Foundation. His research excellence has been recognized with numerous significant awards and scholarships, highlighting his impactful contributions to AI and technology. He currently holds a leadership position as a Co-chair at the Northeastern Educational Research Association, Inc., based in Boston, Massachusetts. Krishna is also an active member of the International Society for Technology in Education (ISTE) and the International Association of Engineers – Hong Kong. Beyond his research and professional activities, Krishna is deeply committed to mentoring, particularly focused on supporting first-generation, underrepresented, and marginalized students in STEM fields. His dedication to fostering inclusivity and diversity within the tech community underscores his commitment to using his expertise to make a positive societal impact. Krishna holds a master's degree in Data Science from the University of Massachusetts Amherst.

About the Technical Reviewer

 **Seshagirirao Lekkala** is a seasoned Cloud and Network Security expert with 16 years of experience in software design and development for the telecommunications sector. Renowned for his expertise in engineering highly scalable, distributed networking solutions tailored for Cloud and AI technologies, his strategic insight and architectural ingenuity have been critical in generating multibillion-dollar revenue for industry giants. His commitment extends beyond technical mastery; he actively fosters the professional development of emerging talent through mentorship and contributes to the industry's body of knowledge through his digital articles, thereby affirming his stature as a powerful influencer and respected authority in Network Security.

Acknowledgments

Gratitude to the pioneers of privacy and AI security. Your insights have illuminated the path in *Privacy in the Age of Innovation*. Your dedication to safeguarding information is instrumental in shaping a secure digital future.

CHAPTER 1

Introduction

The world around us is changing rapidly due to artificial intelligence (AI), which has profound effects on data security, data privacy, and information security. AI-driven solutions are being used more and more to safeguard sensitive data, identify and stop cyberattacks, and adhere to legal requirements. AI does, however, also bring up new privacy and security issues that need to be carefully considered and addressed.

In this book, the intersection of AI, information security, data privacy, and data security is examined. It looks at the situation as it stands, the main risks and challenges, and new trends. In addition, the book offers helpful advice on how to apply AI for improved privacy and security in a moral and responsible way.

The pursuit of knowledge in AI necessitates the stewardship of privacy. Let us, as scientists, be the guardians of innovation, constructing a narrative in which data security is the cornerstone of technological advancement.

1.1 The Intersection of AI, Information Security, Data Privacy, and Data Security

AI is already being applied in several ways to improve data security, data privacy, and information security. AI-driven security solutions, for instance, are able to identify and address cyber threats faster and more efficiently than conventional techniques. AI can also be used to secure sensitive data by detecting and blocking unwanted access, as well as encrypting it while it's in transit and at rest. Additionally, by recognizing and controlling personal data and giving individuals control over it, AI can be used to abide by data privacy laws.

1.2 Outline of the Book

Chapter 1: Introduction

Embark on a journey to understand the intricate relationship between artificial intelligence (AI) and its impact on information security, data privacy, and data security. This chapter sets the stage by defining the scope and objectives of this comprehensive guide, preparing you to delve into the world of AI-powered security solutions.

Subtopics

- The Intersection of AI, Information Security, Data Privacy, and Data Security

- Outline of the Book

- Target Audiences/Readers

Chapter 2: Understanding AI and Ethics

Before exploring the applications of AI in security and privacy, it's crucial to establish a solid foundation in the fundamentals of AI, machine learning, and deep learning. This chapter provides an overview of these concepts, equipping you with the necessary knowledge to comprehend the power of AI.

Subtopics

- Fundamentals of AI, Machine Learning, and Deep Learning

- The Ethics of AI in Privacy and Security

Chapter 3: Information Security and Data Privacy Landscape

Delve into the current state of information security and data privacy, examining the pervasiveness of cyber threats and the evolving landscape of privacy regulations. This chapter highlights the key challenges and risks that organizations face in protecting their sensitive data.

Subtopics

- The Current State of Information Security and Data Privacy

- Key Challenges and Risks

Chapter 4: AI for Threat Detection and Prevention

Discover how AI is revolutionizing threat detection and prevention, empowering organizations to stay ahead of evolving cyberattacks. This chapter showcases AI-driven cybersecurity solutions and real-world case studies, demonstrating the effectiveness of AI in safeguarding sensitive information.

Subtopics

- How AI Can Bolster Threat Detection and Mitigation

- AI-Driven Cybersecurity Solutions

- Case Studies of AI in Action

Chapter 5: Privacy-Preserving AI Techniques

Explore the techniques that enable organizations to leverage AI while preserving individual privacy. This chapter delves into homomorphic encryption, differential privacy, and secure multiparty computation, providing insights into protecting confidential data during AI applications.

Subtopics

- Techniques for Preserving Privacy While Using AI

- Homomorphic Encryption, Differential Privacy, and Secure Multiparty Computation

- Implementing Federated Learning for Data Privacy

Chapter 6: Data Protection and Compliance

Navigate the complex world of data protection and compliance, examining the regulations and standards (e.g., GDPR, CCPA) that govern the handling of personal data. This chapter explores how AI can facilitate compliance with these regulations, ensuring the protection of individual privacy.

Subtopics

- Regulations and Standards (e.g., GDPR, CCPA) Related to Data Privacy and Security

- AI's Role in Achieving and Maintaining Compliance

Chapter 7: Securing AI Models

Understand the importance of securing AI models to maintain the integrity and reliability of AI-powered systems. This chapter outlines best practices for model explainability, fairness, and deployment security, emphasizing the need to protect AI models from potential vulnerabilities.

Subtopics

- Best Practices for Securing AI Models

- Model Explainability and Fairness

- Model Deployment Security

Chapter 8: Case Studies

Gain valuable insights from real-world examples of AI successfully enhancing information security and data privacy. This chapter presents case studies that showcase the practical applications of AI in various industries, offering valuable lessons learned from these implementations.

Subtopics

- Real-World Examples of AI Enhancing Information Security and Data Privacy

- Success Stories and Lessons learned

Chapter 9: AI in Data Privacy and Ethics

Delve into the ethical considerations surrounding the use of AI in data privacy, examining the potential impact of AI on individual privacy rights. This chapter discusses the role of AI ethics in data handling and privacy protection, ensuring that AI is used responsibly and ethically.

Subtopics

- The Ethical Considerations of AI in Privacy

- The Role of AI Ethics in Data Handling and Privacy Protection

Chapter 10: AI and Data Security

Explore AI-driven data security strategies that empower organizations to safeguard their sensitive data from unauthorized access, use, disclosure, disruption, modification, or destruction. This chapter delves into adversarial machine learning and emerging trends in AI and data security, providing insights into future advancements in security techniques.

Subtopics

- AI-Driven Data Security Strategies

- Adversarial Machine Learning in Security

- Emerging Trends in AI and Data Security

Chapter 11: Balance Between Security and Privacy

Strike a delicate balance between security and individual privacy, examining the trade-offs and challenges involved in protecting sensitive data while respecting individual rights. This chapter guides you through the complexities of navigating the regulatory landscape, ensuring compliance and upholding ethical principles.

Subtopics

- Exploring the Trade-Off Between Security and Individual Privacy

- Navigating the Regulatory landscape

Chapter 12: Best Practices and Recommendations

Equip yourself with practical guidelines for organizations to implement AI for enhanced security and privacy. This chapter provides actionable recommendations for responsible and ethical AI practices, empowering you to make informed decisions in your organization.

Subtopics

- Guidelines for Organizations to Implement AI for Enhanced Security and Privacy

- Ensuring Responsible and Ethical AI Practices

Chapter 13: Future Trends and Challenges

Stay ahead of the curve by exploring emerging trends in AI, information security, and data privacy. This chapter identifies potential challenges and opportunities that lie ahead, providing insights into the future of AI-powered security and privacy solutions.

Subtopics

- Emerging Trends in AI, Information Security, and Data Privacy

- Predictions and Potential Challenges

Conclusion

Summarize the key takeaways from this comprehensive guide, consolidating your understanding of AI's impact on information security, data privacy, and data security. This chapter concludes with a vision for the future of AI in this domain, emphasizing the importance of balancing innovation and privacy.

Subtopic

- Summary of the book

Additional Resources

Expand your knowledge and stay up-to-date with the latest developments by accessing a curated collection of references, further reading, and useful websites, tools, and organizations. This chapter provides a comprehensive resource list to continue your learning journey.

Subtopics

- Summarizing Key Takeaways

- The Future of AI, Information Security, Data Privacy, and Data Security with a Focus on Privacy

Glossary: Key Terms and Idea Definitions

Enhance your understanding of the terminology used throughout this book by referencing a comprehensive glossary. This chapter provides clear definitions of key terms and concepts, ensuring that you grasp the nuances of AI-powered security and privacy.

Subtopic

- Definitions of Key Terms and Concepts

1.3 Target Audiences/Readers

This book will give readers a thorough understanding of the intersection of artificial intelligence, information security, data privacy, and data security. It will also offer practical advice on how to use AI to improve security and privacy while adhering to responsible and ethical practices. Anyone interested in the future of AI and its impact on information security and data protection should read this book.

1.3.1 Who Can Read This Book?

This book is intended for a wide range of readers, including

- Experts in information security
- Experts in data privacy
- AI professionals
- Business executives
- Policymakers
- Teachers and students
- Academic experts

CHAPTER 2

Understanding AI and Ethics

Artificial intelligence (AI) is a wide and dynamic field that includes numerous subfields, each of which contributes to its transformational potential. Understanding the principles of AI is essential, particularly in the context of improving information security, data privacy, and data security. This chapter digs into the AI building blocks, digging into machine learning (ML) and deep learning (DL) to uncover their roles in achieving a delicate balance between innovation and privacy.

2.1 Fundamentals of AI, Machine Learning, and Deep Learning

2.1.1 Defining Artificial Intelligence

At its core, artificial intelligence refers to the development of computer systems capable of doing tasks that would ordinarily need human intelligence. These activities span from problem-solving and speech recognition to visual perception and language translation. AI aspires to emulate human cognitive functions, allowing machines to learn from experience, adapt to new knowledge, and execute tasks autonomously.

© Ranadeep Reddy Palle, Krishna Chaitanya Rao Kathala 2024
R. R. Palle and K. C. R. Kathala, *Privacy in the Age of Innovation*,
https://doi.org/10.1007/979-8-8688-0461-8_2

In the field of information security and data privacy, AI functions as a catalyst for innovation by offering intelligent, adaptive systems that improve the ability to detect threats, foresee weaknesses, and efficiently respond to cyber incidents.

2.1.2 The Evolution of AI: From Rule-Based Systems to Machine Learning

AI has progressed from rule-based systems with explicit instructions to a more sophisticated era of machine learning. While rule-based systems excel at certain, well-defined tasks, machine learning allows computers to learn and improve without being explicitly programmed.

2.1.3 Unveiling the Power of Machine Learning

1.   The Basics of Machine Learning: Machine learning is a branch of AI that focuses on developing algorithms that can learn from data. It entails the creation of models capable of detecting patterns, making predictions, and improving their performance over time. Three types of machine learning algorithms exist: reinforcement learning, supervised learning, and unsupervised learning.

2.   Predictive Analytics and Machine Learning in Security: Machine learning (ML) and predictive analytics have emerged as effective threat detection and mitigation tools in the cybersecurity field. From intrusion detection to malware analysis, machine learning algorithms sift through massive amounts of data to detect patterns that indicate potential threats. Intrusion detection systems use machine

learning to detect anomalous behavior within networks and quickly flag suspicious activities for further investigation. Similarly, in malware analysis, machine learning algorithms can quickly identify and classify malicious software, allowing security teams to proactively defend against evolving cyber threats. These applications demonstrate ML's critical role in strengthening digital defenses against an ever-changing threat landscape.

2.1.4 Delving Deeper: Understanding Deep Learning

1. The Architecture of Deep Learning: Deep learning, a subset of machine learning, introduces neural networks with numerous layers (deep neural networks) to replicate the organization of the human brain. Deep neural networks excel at learning detailed patterns and representations from massive volumes of data, making them particularly useful for applications like picture and speech recognition.

2. Applications of Deep Learning in Security: Deep learning is extremely useful in dealing with complicated, unstructured data in the context of information security. It excels at situations like malware detection, where it can detect subtle patterns that indicate dangerous code. Deep learning models also help in the field of behavioral analytics by learning regular user behavior and detecting anomalies that may indicate a security problem.

2.1.5 Privacy-Preserving Techniques in Machine Learning and Deep Learning

1. Homomorphic Encryption: Machine learning with privacy protection offers solutions that allow data to be used for model training without revealing sensitive information. One such technology is homomorphic encryption, which allows computations to be performed on encrypted data without decrypting it. This ensures that sensitive information remains private even during the model training process, which is critical for data privacy.

2. Federated Learning: Federated learning is another privacy-preserving strategy in which machine learning models are trained across decentralized devices rather than centralized servers. This shared learning method enables models to improve without jeopardizing individual user privacy, making it an appropriate technique for applications where data decentralization is critical.

3. Differential Privacy: Differential privacy is the practice of adding noise to individual data points in order to protect the privacy of individuals within a dataset. This technique assures that the inclusion or deletion of a single data point has no significant impact on the model's overall conclusion, protecting privacy while maintaining the utility of the machine learning model.

4. Zero-Knowledge Proofs: With zero-knowledge proofs, two parties can show one another that they possess a certain piece of information without actually disclosing it. Zero-knowledge proofs can be used to confirm a model's accuracy in the context of machine learning and deep learning without disclosing the training data or the model's underlying architecture.

2.1.6 Ethical Considerations in AI, Machine Learning, and Deep Learning

1. Bias and Fairness: The use of AI, machine learning, and deep learning models raises ethical concerns, specifically around bias and fairness. Biases in training data might be perpetuated in model predictions, resulting in biased results. Addressing biases in the development process and implementing techniques to assure fairness in AI applications are critical.

2. Explainability and Transparency: The "black box" aspect of some advanced machine learning and deep learning models presents explainability and transparency issues. Understanding how these models make specific decisions is crucial, particularly in applications where the outcomes of decisions have major real-world ramifications. Building trust in AI systems requires striking a balance between model complexity and interpretability.

2.1.7 Striking the Right Balance: Innovation and Privacy

1. Leveraging AI for Security Innovation: The combination of AI, machine learning, and deep learning technologies unleashes hitherto unimaginable potential in the field of information security. These technologies, ranging from predictive analytics to behavioral analysis, enable firms to address security concerns proactively, anticipate emerging threats, and respond quickly to limit risks.

2. Privacy-First AI Development: As organizations use AI to improve security, it is critical that they prioritize privacy. This includes incorporating privacy concerns into the development lifecycle, implementing privacy-protection techniques, and ensuring compliance with data protection regulations. To strike the right balance between innovation and privacy, a comprehensive approach that incorporates ethical considerations, open practices, and accountability in AI deployment is required.

2.1.8 Case Studies: AI and Privacy in Action

1. Healthcare: Balancing Health Data Innovation and Patient Privacy: AI-driven technologies are critical in advancing diagnostics, treatment optimization, and personalized medicine in the healthcare

sector. The sensitive nature of health data, on the other hand, necessitates a meticulous approach to privacy. Techniques like federated learning allow healthcare institutions to collaborate without jeopardizing patient privacy, fostering innovation while maintaining confidentiality.

2. Finance: Enhancing Security in Financial Transactions: The financial industry relies on AI and machine learning to detect fraud, manage risk, and authenticate customers. Financial institutions can analyze transaction data without exposing sensitive information by using privacy-preserving techniques such as homomorphic encryption. This ensures strong security measures without jeopardizing individual financial transactions' privacy.

2.2 The Ethics of AI in Privacy and Security

2.2.1 The Intersection of Innovation and Ethics

The rapid integration of AI into various aspects of our lives, particularly in the realms of privacy and security, necessitates a thorough understanding of the ethical implications. While AI provides unprecedented opportunities for innovation, it also introduces ethical challenges that must be carefully considered. Exploring the ethical dimensions of AI is critical in this context to ensure that technological advances align with values such as privacy, fairness, transparency, and accountability.

2.2.2 Bias and Fairness: Addressing Ethical Quandaries

The potential for bias is one of the most important ethical considerations in the deployment of AI in privacy and security. Bias can enter AI models via biased training data, resulting in discriminatory outcomes. Biased algorithms in security applications may result in disproportionate targeting of specific groups or individuals, undermining the fairness and equity that should be inherent in security measures.

Addressing bias requires a proactive approach that includes carefully scrutinizing training datasets, being aware of potential biases, and implementing strategies to mitigate and correct bias in AI algorithms. Ethical AI development necessitates ongoing efforts to create models that are not only effective in security applications, but also objective and fair in their decision-making processes.

2.2.3 Explainability and Transparency: Fostering Trust in AI Systems

The lack of transparency and explainability in AI systems poses ethical challenges as they become more complex. Understanding how AI makes specific decisions in the realms of privacy and security is critical for fostering trust among users, stakeholders, and the general public. Explainability is especially important in scenarios where AI-driven decisions have real-world implications, such as security threat assessments or access control systems.

To ensure transparency in AI models, methods for explaining the rationale behind decisions must be developed, making the decision-making process understandable to non-experts. It is critical to strike a balance between model complexity and interpretability in order to implement effective security measures while maintaining transparency to address ethical concerns.

2.2.4 Accountability in AI: Navigating the Complex Web

The use of artificial intelligence in privacy and security creates a complex web of accountability. As AI systems assume decision-making roles, concerns about responsibility and accountability become more prominent. Establishing clear lines of accountability is critical in scenarios where AI is used to assess security threats or make privacy-related decisions.

Ethical AI development entails developing accountability frameworks, clarifying the roles and responsibilities of human operators, and establishing mechanisms for oversight and redress in the event of incorrect decisions. Accountability is not only a legal requirement, but also a fundamental ethical principle to protect against potential misuse or unintended consequences of artificial intelligence in privacy and security applications.

2.2.5 Striking the Right Balance: Ethical Decision-Making in Security

Balancing innovation and ethical concerns is a constant problem, especially in the fast growing field of artificial intelligence. Ethical decision-making in security applications requires a holistic approach that considers not just the technical aspects of AI but also its social, cultural, and legal ramifications.

Organizations must take a proactive approach, incorporating ethics into the development process and ensuring that security measures are consistent with societal values. To strike the right balance, technologists, ethicists, policymakers, and end users must work together to develop a framework that promotes innovation while upholding ethical standards.

1. The Human Element: Ensuring Human Oversight and Intervention: While AI helps to improve information security and privacy, the human element remains indispensable. Human oversight and intervention mechanisms are included in ethical AI deployment, recognizing that AI systems are designed and operated by humans. Humans are in charge of defining objectives, establishing ethical boundaries, and interpreting AI model outputs. When AI encounters ambiguous scenarios, ethical quandaries, or novel situations that were not part of its training data, human oversight becomes critical. Ethical AI frameworks should emphasize the value of human judgment, ensuring that decisions with ethical implications are not fully delegated to AI systems but instead involve human deliberation and accountability.

2. Privacy-Preserving AI and Ethical Considerations: Privacy-preserving AI approaches are critical in the quest to improve information security and data privacy. However, the application of these tools raises a new set of ethical concerns. The desire for data-driven insights must be balanced with the requirement to protect individual privacy. To ensure that privacy-preserving strategies are ethical, consumers must be informed about the methods used to secure their data. Consent, transparency, and user empowerment have all become essential components of ethical privacy-preserving AI, ensuring that users are informed participants in the use of their data for AI applications.

3. Ethical Decision-Making Frameworks for AI
    in Security: Creating ethical decision-making
    frameworks tailored to artificial intelligence in
    security is critical for directing responsible and
    transparent activities. Fairness, accountability,
    transparency, and diversity should be incorporated
    into these frameworks. Furthermore, they should
    be adaptive, given the ever-changing nature of both
    technology and ethical norms.

2.2.6 Case Studies: Navigating Ethical Challenges in AI Security Applications

1. Facial Recognition Technology: Facial recognition
    technology, while useful in security applications,
    raises serious ethical concerns. Privacy advocates
    emphasize the possibility of constant surveillance
    and restrictions on individual freedoms.
    Furthermore, these algorithms have been
    demonstrated to exhibit racial and gender biases,
    resulting in false positives and discriminatory
    outcomes. Furthermore, governments and
    corporations may misuse facial recognition data for
    mass surveillance or social control, raising serious
    concerns. Case studies that investigate these issues
    will be critical in developing ethical frameworks
    and regulations to ensure the responsible
    implementation of AI security applications.

2. Predictive Policing: Ethical Implications: The
    employment of artificial intelligence in predictive
    policing is a tricky ethical issue. While predictive

algorithms promise to improve law enforcement by identifying future crime hotspots, ethical concerns about prejudice, transparency, and the possibility for perpetuating existing inequities necessitate careful consideration. Case studies in predictive policing offer insight on the ethical problems confronting law enforcement agencies, as well as the importance of transparency, impartiality, and accountability in the implementation of AI-driven security measures.

3. Social Media Monitoring: Privacy and Freedom of Expression: Monitoring social media platforms for security purposes raises ethical concerns about privacy and freedom of expression. Case studies in social media surveillance demonstrate the delicate balance between safeguarding public safety and respecting individual privacy rights. In this context, ethical decision-making entails establishing clear boundaries, assessing the potential influence on democratic principles, and putting safeguards in place to prevent overreach.

2.2.7 Navigating the Ethical Landscape of AI in Privacy and Security

Eventually, Chapter 2 explored the ethics of AI in the context of improving information security, data privacy, and data security. As artificial intelligence (AI) technology becomes more integrated into security systems, ethical considerations are critical to ensuring responsible and equitable deployment. From tackling bias and increasing transparency to developing accountability frameworks and managing difficult cases, there is something for everyone.

CHAPTER 3

Information Security and Data Privacy Landscape

The information security and data privacy environment is a fluid terrain influenced by technology breakthroughs, new threats, and legislative frameworks. In an era dominated by artificial intelligence (AI), the environment is witnessing a dramatic transformation in how enterprises safeguard critical information. Striking a careful balance between innovation and privacy is critical. This landscape is distinguished by the ongoing development of ethical frameworks, the incorporation of privacy-preserving AI approaches, and a greater emphasis on transparency and responsibility. As enterprises negotiate this complicated landscape, they must adapt to the ever-changing nature of cyber threats while respecting ethical standards and protecting individual privacy rights.

© Ranadeep Reddy Palle, Krishna Chaitanya Rao Kathala 2024
R. R. Palle and K. C. R. Kathala, *Privacy in the Age of Innovation*,
https://doi.org/10.1007/979-8-8688-0461-8_3

3.1 The Current State of Information Security and Data Privacy

In the ever-changing context of information security and data privacy, a comprehensive grasp of the current state is critical. This section digs into the multiple components that shape the current landscape, examining the problems, advancements, and upcoming trends that define the delicate tapestry of information security and data privacy.

3.1.1 The Evolving Threat Landscape

A dynamic and sophisticated threat environment characterizes today's information security landscape. Cyberattackers are constantly updating their strategies, techniques, and procedures in order to exploit weaknesses and avoid traditional security measures. The threat landscape has grown in both magnitude and complexity, with ransomware attacks affecting vital infrastructure and sophisticated phishing operations targeting individuals.

In order to protect against these risks, organizations must take a proactive and flexible approach. To supplement traditional security measures, the present condition of information security demands ongoing attention, threat intelligence exchange, and the use of sophisticated technologies such as AI.

3.1.2 Regulatory Frameworks and Compliance

In response to growing concerns about data breaches and privacy violations, regulatory frameworks around the world have been significantly expanded and revised. Initiatives such as the European Union's General Data Protection Regulation (GDPR) and the United States' California Consumer Privacy Act (CCPA) show the global trend toward higher data protection requirements.

Compliance with these standards has become an essential component of today's data privacy landscape. Organizations must traverse a complicated web of legal regulations to ensure that their data handling procedures are consistent with the principles of transparency, consent, and data subjects' rights. Maintaining confidence with consumers and stakeholders has grown increasingly dependent on the confluence of compliance and effective security measures.

3.1.3 Rise of Privacy-Preserving Technologies

With rising concerns about data privacy, there has been a significant increase in the adoption of privacy-protecting technologies. Techniques like homomorphic encryption, federated learning, and differential privacy are gaining traction as corporations strive to harness new technology while protecting sensitive data.

These privacy-preserving technologies signify a paradigm shift toward a more responsible and ethical approach to data processing, aligning with the increasing emphasis on user privacy and data protection.

3.1.4 Balancing Act: Innovation Versus Privacy

The current condition of information security and data privacy reflects a continual balancing act between the requirement to safeguard individual privacy and the imperative to innovate. The growing incorporation of AI into security systems brings unprecedented threat detection, predictive analytics, and automation possibilities. This innovation, however, must be balanced by a thorough understanding of ethical factors such as bias mitigation, transparency, and responsibility.

To achieve the proper balance, firms must take a comprehensive approach that includes not only technology developments but also robust rules, employee training, and a privacy-centric organizational culture.

The issue is to capitalize on the benefits of innovation while maintaining ethical standards and ensuring that privacy remains a fundamental right in the digital era.

3.1.5 Increasing Awareness and Education

In today's environment, there is a rising realization of the significance of cybersecurity knowledge and education. End users, employees, and decision-makers are all becoming increasingly aware of the risks associated with poor security policies and the potential implications of data breaches.

Organizations are investing in comprehensive training programs to improve the cybersecurity literacy of their personnel. From spotting phishing efforts to comprehending the ramifications of data sharing, education is a critical component of reinforcing the human aspect against new threats. Increased awareness develops a culture of shared responsibility in which individuals actively contribute to the organization's overall security posture.

3.1.6 Cloud Security and the Remote Work Paradigm

The growing usage of cloud computing has transformed how businesses manage data and information technology infrastructure. Although cloud services provide scalability, flexibility, and cost-effectiveness, they also pose new security concerns. The current condition of information security necessitates a thorough examination of cloud security measures such as data encryption, access controls, and secure configuration standards.

Furthermore, the paradigm shift toward remote work has been hastened by global events, emphasizing the significance of securing remote access and endpoints. To accommodate a distributed workforce, organizations must adjust their security strategy, highlighting the importance of resilient virtual private networks (VPNs), multi-factor authentication, and secure collaboration technologies.

3.1.7 Incident Response and Resilience

Because security incidents are unavoidable, companies are emphasizing incident response and resilience. The present level of information security recognizes the importance of having a proactive and well-defined incident response strategy in place to mitigate the consequences of a security breach.

Resilience is not just effectively responding to incidents, but also learning from them in order to improve future security measures. To test and improve their incident response capabilities, organizations are investing in continuous improvement procedures, threat intelligence exchanges, and simulations.

Navigating the Complex Terrain

To summarize, the current state of information security and data privacy is a complex and dynamic landscape affected by technology advances, legislative frameworks, and increasing threat landscapes. Organizations must negotiate this complex landscape with a complete approach that incorporates technology breakthroughs, regulatory compliance, privacy-preserving technologies, and an ethical commitment.

As the digital landscape evolves, it will be vital to maintain a proactive approach to cybersecurity, privacy education, and resilient incident response. The delicate balance between innovation and privacy necessitates continual awareness, agility, and a commitment to ethical principles, ensuring that enterprises effectively traverse the intricacies of today's information security and data privacy landscape.

3.2 Key Challenges and Risks

Navigating the complex terrain of information security and data privacy requires not just comprehending the current status but also recognizing the key problems and threats that organizations face. This section investigates the multifaceted threats to data integrity, information confidentiality, and the overall security posture of companies in the digital age.

3.2.1 Cybersecurity Threats: A Persistent Menace

The constant and evolving nature of cybersecurity threats is the most significant challenge in the information security landscape. Cyber enemies use increasingly complex approaches to access networks and compromise critical data, including advanced technology, social engineering, and exploit techniques. Threats range from simple malware and ransomware to complicated Advanced Persistent Threats (APTs) that can go undiscovered for long periods of time.

Rapidly developing attack vectors, such as zero-day exploits and fileless malware, put standard security measures to the test, necessitating the implementation of enhanced threat detection and response techniques. To effectively combat attacks, the dynamic nature of these threats necessitates ongoing awareness, threat intelligence sharing, and adaptive cybersecurity methods.

3.2.2 Insider Threats: The Human Element

While external threats get the most attention, insider risks pose a major threat to information security and data privacy. Employees, contractors, or partners acting maliciously or inadvertently can jeopardize critical data or intellectual property. Insider risks can take the shape of intentional data theft, careless management of sensitive information, or unknowing participation in phishing attempts.

Addressing insider threats necessitates a delicate balance of confidence and monitoring. To detect abnormal activity, organizations must have strong access restrictions, personnel training programs, and behavioral analytics. Building a security awareness culture and instilling a sense of responsibility in employees is critical for mitigating the dangers connected with insider threats.

3.2.3 Regulatory Compliance Burden and Complexity of Privacy Regulations

While regulatory frameworks might help improve data privacy, the compliance cost they create can be a substantial challenge for businesses. Navigating a complex network of regulations, each with its own set of requirements, may be time-consuming and error-prone. The General Data Protection Regulation (GDPR), the Health Insurance Portability and Accountability Act (HIPAA), and other regional or industry-specific requirements all require strict adherence to data security standards.

Noncompliance not only exposes firms to legal ramifications, but it also erodes consumer and stakeholder trust. To maintain long-term compliance, it is necessary to build efficient processes, conduct regular audits, and stay current on changing regulatory environments.

The broad and ever-changing environment of privacy legislation presents a significant problem for enterprises wanting to remain compliant. Managing the complexities of global, regional, and industry-specific regulations necessitates a thorough awareness of legal standards, data processing principles, and reporting requirements.

The constant growth of privacy regulations adds to the complication. Keeping up with modifications, updates, and new legislation necessitates a strong commitment to ongoing learning and adaptability. Companies must invest in legal expertise, conduct regular privacy effect assessments, and put in place adaptable frameworks that can adapt to changing regulatory environments.

3.2.4 Integration of Privacy-Preserving Technologies

While privacy-preserving technologies provide intriguing solutions, their integration presents practical and adoption hurdles. Organizations must deal with the complexities of integrating technologies such as homomorphic encryption, federated learning, and differential privacy into existing infrastructures. The learning curve, potential performance consequences, and compatibility difficulties may prevent widespread use.

Furthermore, the ethical questions surrounding the usage of these technologies add another layer of complication. Striking the correct balance between privacy and functionality necessitates careful planning, investment in staff training, and a commitment to addressing ethical considerations related with the deployment of new privacy-preserving technology.

3.2.5 Cloud Security Concerns

The shift to cloud computing brings both benefits and hazards to information security. While cloud services provide scalability and flexibility, the shared responsibility model demands enterprises to actively manage their security within the cloud environment. Concerns about data breaches, improper access controls, and poor encryption techniques are common in the cloud security landscape.

Securing cloud systems demands a complete approach that includes encryption technologies, safe configuration practices, and effective access restrictions. Organizations must also address the problems associated with multi-cloud setups and guarantee that third-party cloud service providers comply to high security standards.

3.2.6 Advanced Persistent Threats (APTs): Stealthy Adversaries

In the information security landscape, Advanced Persistent Threats (APTs) pose a serious challenge. APTs are complex, long-term campaigns conducted by well-funded, highly competent threat actors. These adversaries operate with extreme stealth, frequently going undiscovered for extended periods of time, allowing them to steal important data or disrupt operations.

Advanced threat intelligence, constant monitoring, and proactive defense techniques are required for detecting and neutralizing APTs. To stay updated about developing APT methods, organizations must invest in technologies that can detect subtle symptoms of compromise, perform extensive incident response planning, and interact with the larger cybersecurity community.

3.2.7 Skills Shortage in Cybersecurity

The shortage of experienced individuals in the cybersecurity industry is a serious issue that enterprises all over the world are dealing with. The demand for skilled cybersecurity personnel has increased as cyberattacks become more complex. Threat analysis, incident response, and safe software development are all suffering from a labor scarcity.

To close the skills gap, organizations must execute complete workforce development strategies that include training programs, certifications, and talent acquisition campaigns. Collaboration with educational institutions, industry associations, and the cybersecurity community can help to establish a strong and trained workforce capable of addressing emerging information security threats.

Navigating Complexity with Vigilance

Consequently, the world of information security and data privacy presents a plethora of problems and hazards that enterprises must traverse with diligence and strategic thinking. The situation necessitates a comprehensive and flexible approach, from the ongoing danger of cyberattacks and the challenges of regulatory compliance to the integration of sophisticated technologies and the scarcity of cybersecurity expertise.

Addressing these difficulties necessitates a commitment to lifelong learning, investment in technical advances, collaboration with peers in the industry and regulatory agencies, and a proactive approach to developing a resilient cybersecurity posture. Recognizing key obstacles and hazards as companies negotiate this complicated terrain provides as a foundation for informed decision-making and the creation of successful methods to manage sensitive information and preserve data privacy.

CHAPTER 4

AI for Threat Detection and Prevention

4.1 How AI Can Bolster Threat Detection and Mitigation

In the never-ending arms race between cyberattackers and defenders, artificial intelligence (AI) emerges as a powerful force for enhancing threat identification and mitigation. This section investigates AI's revolutionary capabilities in bolstering security measures and improving the ability to identify, respond to, and avoid cyber threats.

4.1.1 Advanced Threat Detection with Machine Learning

Traditional signature-based detection methods struggle to keep up with the constantly changing threat landscape. Static signature lists necessitate frequent updates, making systems vulnerable to novel attacks. Furthermore, these methods are primarily reactive, relying on predefined

© Ranadeep Reddy Palle, Krishna Chaitanya Rao Kathala 2024
R. R. Palle and K. C. R. Kathala, *Privacy in the Age of Innovation*,
https://doi.org/10.1007/979-8-8688-0461-8_4

patterns to detect threats. In contrast, machine learning takes a proactive approach. By continuously learning from new data, machine learning algorithms can detect subtle anomalies and emerging threats that traditional signature-based methods may miss.

Machine learning (ML), a subtype of AI, is on the cutting edge of threat detection. Traditional signature-based methods detect known threats but fall short of detecting novel and sophisticated attacks. Machine learning algorithms, on the other hand, excel in analyzing massive datasets to identify patterns, abnormalities, and indicators of compromise.

ML models grow adept at spotting deviations from typical behavior through continual learning from historical data. Using this adaptive method, organizations can detect emerging threats, zero-day exploits, and polymorphic malware that may avoid traditional detection mechanisms. Threat detection powered by machine learning provides a proactive defense, helping enterprises to remain ahead of cyber threats.

4.1.2 Predictive Analytics: Anticipating Threats Before They Manifest

AI-driven predictive analytics ushers in a paradigm change from reactive to proactive threat mitigation. Predictive analytics allows firms to anticipate possible dangers by studying previous data and detecting patterns. This proactive strategy enables security professionals to take preventative steps, narrowing the window of vulnerability and decreasing the impact of cyberattacks.

Predictive analytics powered by AI can estimate the likelihood of specific attack vectors, identify weak assets, and prioritize security activities depending on threat severity. This strategic application of artificial intelligence not only improves threat detection but also helps to a more efficient and resource-effective security posture.

4.1.3 Behavioral Analytics: Understanding the Human Element

Traditional security measures face a difficult task in understanding the intricacies of human behavior in the digital domain. AI-driven behavioral analytics transforms threat detection by creating baselines for regular user behavior and detecting deviations that may indicate malicious activity.

Behavioral analytics can detect anomalies such as unusual login timings, access patterns, or data transfer volumes by continuously learning from user interactions. This human-centric approach to threat detection adds another level of sophistication, helping firms to discover insider threats, compromised accounts, and other security issues that rule-based systems may miss.

4.1.4 Anomaly Detection: Uncovering Stealthy Threats

An important use of AI is anomaly detection, which includes identifying deviations from predefined baselines. This approach is particularly useful at detecting covert and evasive attacks in the context of threat detection. Anomaly detection can generate alarms based on variations from regular system behavior, whether it's an insider threat seeking illegal access or a subtle reconnaissance phase by a potential attacker.

Large datasets can be analyzed by AI algorithms to establish baseline behavior for networks, endpoints, and user activities. Anomaly detection systems generate alerts when deviations occur, suggesting possibly harmful behavior. This feature is critical for detecting threats that do not follow the conventional patterns associated with established attack pathways.

4.1.5 Natural Language Processing (NLP): Enhancing Contextual Understanding

Natural language processing (NLP) is becoming increasingly important as cyber threats expand to encompass more sophisticated social engineering and phishing attempts. NLP helps AI systems to comprehend and evaluate human language, hence improving the contextual understanding of security occurrences. This is especially important in cases where threats are embedded in emails, messages, or other forms of communication.

AI-driven natural language processing (NLP) can detect tiny subtleties in language, detecting phishing attempts, social engineering strategies, and communication patterns suggestive of malevolent intent. Organizations can strengthen their defenses against threats that use human contact as a vector by incorporating NLP into threat detection systems.

4.1.6 Adversarial Machine Learning: A Cat-and-Mouse Game

While AI facilitates defenders, cyberattackers use it to craft more sophisticated and evasive attacks. Adversarial machine learning refers to attempts by attackers to manipulate or deceive AI models by giving them false or harmful data. This cat-and-mouse game puts AI-driven threat detection technologies to the test.

To withstand manipulation, defenders must anticipate adversarial techniques and constantly update their AI models. Model ensembling, dynamic retraining, and the utilization of different datasets all contribute to the development of more resilient AI systems capable of adapting to developing adversarial methods.

4.1.7 Automation and Orchestration: Swift Response to Threats

AI not only improves threat detection but also allows for faster and more automated responses to security problems. Organizations can use automation and orchestration to streamline the response cycle, from alarm triage to containment and remediation. AI-driven systems can perform programmed reactions automatically based on the intensity and nature of detected threats.

Automation shortens response times, decreases the possibility of human mistake, and provides a uniform and coordinated response to security issues. Organizations can efficiently handle the growing number and complexity of cyber threats by orchestrating responses using AI-driven processes.

4.1.8 Integration with Threat Intelligence Feeds

When AI is combined with threat intelligence feeds, its efficacy in threat identification is enhanced. These feeds deliver real-time updates on emerging threats, known adversaries, and the most recent tactics, strategies, and procedures (TTPs). AI systems can examine this intelligence and correlate it with internal data to improve threat detection accuracy and relevance.

Organizations may keep ahead of the ever-changing danger landscape by integrating with threat intelligence streams. AI-driven threat detection becomes more contextually rich and flexible by integrating external intelligence sources, matching with the dynamic nature of cyber threats.

Augmenting Security Defenses with AI

While AI excels at threat detection, its true value comes from its ability to transform security beyond reactive measures. AI-powered systems can analyze threat intelligence and historical data to anticipate potential attacks before they happen. This proactive approach enables security teams to take preventive measures like prioritizing vulnerabilities or isolating compromised systems. This shift from reactive defense to predictive anticipation represents a significant advancement in cybersecurity capabilities, allowing organizations to establish a more resilient and adaptable security posture.

The ever-changing world of cyber threats needs a proactive and informed strategy for security. By leveraging AI capabilities, organizations may not only detect and respond to threats more effectively but also remain ahead of the curve in an environment where cyberattackers are growing more sophisticated. As we negotiate the complex terrain of information security, the role of AI in threat detection emerges as a critical aspect in bolstering digital defenses and minimizing the growing hazards of the digital era.

4.2 AI-Driven Cybersecurity Solutions

In the ever-changing environment of cybersecurity, integrating artificial intelligence (AI) has become a critical component in bolstering defenses against a wide range of attacks. This section digs into the wide range of AI-driven cybersecurity solutions that enable enterprises to detect, respond to, and prevent cyber threats proactively.

4.2.1 Machine Learning-Powered Endpoint Protection

Endpoints, which include PCs, servers, and mobile devices, are popular targets for cybercriminals. Machine learning algorithms are used in AI-driven endpoint protection to continually evaluate endpoint data, discover patterns suggestive of malicious activity, and respond to possible threats autonomously.

These solutions go beyond standard signature-based approaches by learning from the behavior of both normal and malicious processes to adapt to the shifting threat landscape. As a result, the defense system becomes more dynamic and effective, allowing it to quickly recognize and neutralize threats at the point of entry.

4.2.2 Behavioral Analytics for User-Centric Security

AI-driven behavioral analytics is changing how businesses approach user-centric security. Behavioral analytics solutions detect anomalies such as hacked accounts or insider threats by defining baseline behaviors for specific individuals and entities.

As AI systems learn from user interactions, they grow more capable of discriminating between normal and deviant behavior. This method not only improves threat detection but also reduces false positives, allowing enterprises to focus their resources on legitimate security issues.

4.2.3 Network Traffic Analysis with AI

Analyzing network data is a critical component of cybersecurity, and AI introduces a new level of sophistication to the task. AI-driven network traffic analysis entails continuously monitoring and analyzing network data to detect strange trends, anomalies, or potential security concerns.

These systems can detect advanced threats including lateral network movement or malware connecting with command and control servers. AI improves the capacity to quickly discover and respond to potential security breaches by automating the examination of massive volumes of network data.

4.2.4 Automated Threat Hunting and Incident Response

AI-driven cybersecurity solutions automate the previously time-consuming procedures of threat detection and incident response. Threat hunting is the proactive pursuit of potential threats within an organization's environment, and AI augments this process by autonomously evaluating massive datasets for indicators of compromise.

AI-driven incident response solutions offer automatic actions based on specified playbooks. These technologies streamline the response workflow, lowering the time it takes to control and mitigate security incidents, from isolating infected systems to performing repair steps.

4.2.5 Predictive Analysis for Vulnerability Management

The predictive analysis capabilities of AI extend to vulnerability management, in which businesses aim to identify and remedy potential vulnerabilities in their systems before they can be exploited. AI-driven solutions can forecast where vulnerabilities are most likely to be exploited by analyzing historical data, threat intelligence feeds, and system configurations.

Organizations can more efficiently deploy resources and shorten the window of vulnerability by prioritizing remediation operations based on the potential impact and exploitability of vulnerabilities. This predictive technique is in line with a proactive cybersecurity strategy, as it prevents possible attacks before they occur.

4.2.6 AI-Enhanced Phishing Detection

Phishing attacks continue to be a prominent and potent threat in the cybersecurity world. AI-driven phishing detection solutions use machine learning algorithms to scan email content, sender activity, and contextual information to identify phishing attempts.

These technologies go beyond typical email filters, reacting to the growing techniques of cybercriminals. AI-driven phishing detection technologies improve the accuracy of recognizing fraudulent emails and lower the risk of successful phishing assaults by learning from patterns associated with phishing campaigns.

4.2.7 Autonomous Threat Intelligence Platforms

AI-driven threat intelligence solutions process and analyze massive volumes of threat data using machine learning. These systems can collect information from a variety of sources, such as open-source intelligence, dark web forums, and historical incident data. AI-driven solutions give enterprises with real-time insights into emerging threats and developing tactics by continuously learning from this data.

Autonomous threat intelligence solutions enable enterprises to stay ahead of the curve by making informed decisions based on the most up-to-date and relevant threat intelligence. This proactive strategy improves overall cybersecurity resilience.

4.2.8 Continuous Adaptive Risk and Trust Assessment

Continuous Adaptive Risk and Trust Assessment (CARTA) is an approach that uses artificial intelligence (AI) to continuously assess and update security measures based on the changing risk landscape. To dynamically alter security postures, AI algorithms examine data from numerous sources, including user behavior, system configurations, and threat intelligence.

CARTA allows enterprises to transition from a static, perimeter-focused security strategy to a more adaptable, responsive one. Organizations may better align their security measures with the expanding threat landscape and the dynamic nature of modern IT infrastructures by employing AI for continuous assessment.

A Holistic Defense with AI

Ultimately, AI-driven cybersecurity solutions offer a paradigm shift in how businesses defend against cyberattacks. These solutions, which range from machine learning-powered endpoint protection to AI-enhanced cloud security, provide a comprehensive and adaptive defense mechanism.

As the cybersecurity landscape evolves, the use of AI gives enterprises with the agility and efficacy required to keep up with sophisticated adversaries. The adaptability of AI-driven solutions extends beyond threat detection and response to incident response and vulnerability management, enabling enterprises to develop resilient and proactive cybersecurity postures in the face of an ever-changing threat landscape.

4.3 Case Studies of AI in Action

The application of artificial intelligence (AI) in the dynamic landscape of cybersecurity has resulted in transformative effects in threat identification and prevention. Real-world case studies demonstrate the practical impact of AI-driven solutions in countering sophisticated cyberattacks. This section looks at prominent cases where corporations have effectively used AI to improve their cybersecurity posture.

4.3.1 Case Study 1: Darktrace's Autonomous Response at Maersk

The 2017 NotPetya ransomware outbreak, which caused havoc on global companies, notably impacted Maersk, a major shipping and logistics corporation. In the aftermath, Maersk resorted to Darktrace, an AI-driven cybersecurity business, to shore up its defenses.

Darktrace's technology, which is based on unsupervised machine learning, operates on the principles of self-learning and anomaly detection. In the case of Maersk, Darktrace's AI discovered suspicious NotPetya activity before standard security measures could. Notably, Darktrace's autonomous response capacity acted immediately to mitigate the threat, isolating afflicted systems and averting further damage.

This scenario emphasizes AI's agility and proactive behavior in the face of quickly shifting threats. Without predefined signatures, Darktrace's AI autonomously discovered and responded to a sophisticated attack, demonstrating the potential for AI to outperform traditional security solutions in real-time threat scenarios.

4.3.2 Case Study 2: Cylance's AI-Driven Endpoint Protection

Cylance, a cybersecurity company, implemented AI-driven endpoint protection to counter advanced threats for a global pharmaceutical firm. Traditional antivirus solutions often struggle to keep pace with new and evolving malware variants. In this scenario, Cylance's AI-driven approach utilized machine learning models trained on a diverse dataset to identify malicious files based on their characteristics, rather than relying on known signatures.

The pharmaceutical firm experienced a significant reduction in the number of malware incidents, showcasing the efficacy of AI-driven endpoint protection in preventing attacks that would typically evade traditional defenses. The ability to proactively identify and neutralize threats based on behavior and characteristics demonstrated the advantages of AI in mitigating the evolving landscape of cybersecurity threats.

4.3.3 Case Study 3: IBM Watson for Cyber Security at a Financial Institution

A major financial institution deployed IBM Watson for Cyber Security, an AI-driven solution that uses natural language processing and machine learning. The organization had difficulties in promptly reviewing enormous amounts of security data to identify potential risks. The cognitive capabilities of IBM Watson allowed the business to streamline its threat detection and response operations.

IBM Watson's AI systems searched through organized and unstructured data, correlating information to detect patterns indicative of harmful activities. The approach dramatically decreased the time required for threat investigation and allowed security analysts to focus on more strategic areas of cybersecurity. This case study demonstrates how artificial intelligence may improve the efficiency of security operations, allowing firms to traverse the intricacies of today's threat landscape.

4.3.4 Case Study 4: Palo Alto Networks Cortex XDR

Palo Alto Networks deployed Cortex XDR, an AI-driven Extended Detection and Response platform, for a multinational technology company. Faced with difficulties identifying and responding to advanced attacks across its global network, the IT firm needed a system that could provide comprehensive visibility and automatic responses.

Cortex XDR's AI-driven approach correlated threat indicators and identified complicated attack chains by integrating endpoint, network, and cloud data. The platform's machine learning models reacted to new threats in real time, improving the organization's ability to detect and thwart sophisticated attacks. This case study demonstrates how AI-driven XDR platforms may provide complete threat visibility and response capabilities in a variety of scenarios.

4.3.5 Case Study 5: Google's AI for Phishing Detection

Google uses artificial intelligence to improve its email security and phishing detection capabilities. Google's AI-driven method aims to identify and thwart phishing attempts in real time, with millions of users relying on Gmail for communication.

To discover patterns linked with phishing, the AI algorithms assess numerous aspects such as email content, sender behavior, and contextual information. The technology adapts to changing phishing tactics by learning from user interactions. This method has considerably enhanced

Gmail's capacity to recognize and block phishing emails, protecting users against deceptive attempts. The case emphasizes the practical application of AI in combating one of the most prevalent cybersecurity threats— phishing.

4.3.6 Case Study 6: FireEye's Helix Security Platform

The Helix Security Platform from FireEye blends AI and machine learning to improve threat detection and response. A multinational retail firm encountered difficulties in detecting and managing advanced attacks throughout its network. With its AI-driven capabilities, FireEye's Helix offers a holistic solution.

The platform's AI evaluated network and endpoint data in real time, detecting anomalies and potential threats. Machine learning models that are tailored to the corporate environment, reducing false positives and allowing for quick responses to serious risks. The retail corporation's installation of FireEye's Helix resulted in greater threat visibility, reduced response times, and increased overall cybersecurity resilience.

4.3.7 Case Study 7: Symantec's AI-Enhanced Cloud Security

Symantec, a cybersecurity firm, utilized artificial intelligence-driven solutions to improve cloud security for a large technology services provider. As the provider embraced cloud services for its operations, protecting a dynamic and distributed environment became a difficulty.

Machine learning was used to evaluate patterns of user behavior, discover anomalies, and identify potential security threats in Symantec's AI-enhanced cloud protection. The solution was tailored to the cloud infrastructure of the provider, giving real-time insights and automatic reactions to security issues. This study demonstrates how artificial intelligence (AI) may play a critical role in safeguarding cloud settings, in line with the transition toward cloud-centric operations.

Unleashing the Power of AI in Cybersecurity

These case studies clearly demonstrate AI's revolutionary influence in protecting enterprises against a wide range of cyber threats. AI-driven cybersecurity solutions demonstrate a paradigm shift in the efficacy and reactivity of modern security measures, from autonomously responding to ransomware attacks to proactively identifying and combating sophisticated threats.

As organizations struggle with a shifting threat landscape, the use of AI emerges as a strategic priority. AI's ability to learn, adapt, and respond autonomously places it as a strong partner in the never-ending struggle against cyberattackers. These real-world examples demonstrate the viability and efficacy of AI-driven cybersecurity, heralding a new era in which intelligent technologies play a critical role in protecting the digital domain.

CHAPTER 5

Privacy-Preserving AI Techniques

5.1 Techniques for Preserving Privacy While Using AI

Safeguarding sensitive information has become critical in the complex intersection of artificial intelligence (AI) and privacy. As organizations use AI to harvest insights from massive datasets, it is critical that they use strategies that prioritize privacy. This section investigates several ways for protecting privacy while utilizing AI's capabilities.

5.1.1 Homomorphic Encryption: Unlocking Secure Computations

Homomorphic encryption is a key approach for privacy-preserving AI. This cryptographic approach enables computations on encrypted data to be conducted without the requirement for decryption. Homomorphic encryption offers secure model training and inference on encrypted data in the context of machine learning and AI.

Companies can use homomorphic encryption to ensure that sensitive data is encrypted throughout the AI workflow. This strategy is especially useful when working together on data analysis without releasing raw data, establishing a balance between the demand for insights and the requirement to protect privacy.

5.1.2 Federated Learning: Decentralized Model Training

Federated learning decentralizes the model training process, which alleviates privacy concerns associated with centralized data sources. Models are trained across decentralized devices such as smartphones or edge devices without exchanging raw data in this manner. Based on its local data, each device computes a model update, and only the model updates are exchanged with a central server.

This method not only protects user privacy by storing raw data on user devices, but it also allows companies to gain relevant insights from distributed datasets. Federated learning is especially useful in circumstances where data privacy is critical, such as healthcare, where patient data is kept on local devices while contributing to the improvement of a global model.

5.1.3 Differential Privacy: Adding Noise for Privacy Protection

Differential privacy is a technique that adds controlled noise to individual data points to prevent unique persons from being identified in a dataset. Differential privacy is used in the context of AI during data collection and analysis to preserve the privacy of persons who contribute to the dataset.

Differential privacy ensures that the output of an algorithm does not expose information about a specific individual by introducing randomness into the calculation process. This approach provides a strong layer of privacy protection, making it difficult for bad actors to reverse-engineer or de-anonymize specific data points.

5.1.4 Secure Multiparty Computation: Collaborative Data Analysis

Secure multiparty computations (SMPC) allows many parties to compute a function concurrently while keeping their inputs private. In the field of artificial intelligence, SMPC enables businesses or entities to study data collectively without disclosing the raw data itself.

This strategy is useful when multiple entities desire to pool their resources for a common AI goal without exposing sensitive information. Financial firms, for example, can use SMPC to examine transaction data collectively without jeopardizing individual transaction details.

5.1.5 Homomorphic Databases: Securing Query Processing

Homomorphic databases apply the homomorphic encryption idea to databases, enabling secure query processing on encrypted data. Databases can use this technology to execute computations on encrypted data without decrypting it, giving an extra layer of privacy protection for data at rest.

This is especially important in AI applications where queries on sensitive data must be performed without jeopardizing individual records. Homomorphic databases provide secure data analytics and processing while keeping raw information private.

5.1.6 Zero-Knowledge Proofs: Verifying Without Revealing

Zero-knowledge proofs allow one party to demonstrate to another that they have specific information without revealing the information itself. Zero-knowledge proofs can be used in the context of privacy-preserving AI to certify the accuracy of calculations or the possession of specific data without releasing the underlying data itself.

This technique is useful in situations where verification is required but the raw data should be kept private. In a collaborative AI effort, for example, zero-knowledge proofs can be used to demonstrate that each person supplied valid information without revealing the specifics of their particular datasets.

5.1.7 Synthetic Data Generation: Mimicking Real Data Without Exposure

The process of developing artificial datasets that mirror the statistical features of real data without exposing real user information is known as synthetic data production. This method is beneficial when organizations need to share or cooperate on datasets without jeopardizing individual privacy.

Organizations can execute numerous AI operations, such as training models and conducting analysis, without relying on real user data by creating synthetic datasets. This technique reduces the danger of handling sensitive information while yet allowing for important AI applications.

Building a Privacy-Centric AI Landscape

Moreover, the use of AI does not have to come at the expense of individual privacy. These privacy-protection solutions offer a rich tapestry of strategies for enterprises to strike the delicate balance between employing

AI for insights and safeguarding sensitive information. Each methodology, from cryptographic methods like homomorphic encryption to decentralized approaches like federated learning, helps to build a privacy-centric AI landscape.

As enterprises expand their usage of AI applications, adopting these privacy-preserving strategies becomes critical not only for legal compliance but also as a commitment to ethical and responsible data use. The developing environment of privacy-preserving AI represents a paradigm change toward a future in which innovation and secrecy coexist, allowing AI's transformational potential to be realized without jeopardizing individual privacy.

5.2 Homomorphic Encryption, Differential Privacy, and Secure Multiparty Computation

In order to harness the power of artificial intelligence (AI), enterprises must maintain privacy while collecting important insights from sensitive data. This section digs into three crucial privacy-preserving approaches— homomorphic encryption, differential privacy, and secure multiparty computation—and examines how they help to achieving a delicate balance between data value and individual privacy.

5.2.1 Homomorphic Encryption: Preserving Confidentiality in Computation

Homomorphic encryption is a cryptographic innovation that allows secure computations to be performed on encrypted data. The capacity of homomorphic encryption to execute operations on data without the requirement for decryption is its essential concept. In the context of artificial intelligence, this translates to the ability to train machine learning models and execute analysis on encrypted datasets while protecting the confidentiality of the underlying data.

Consider the following scenario: a healthcare institution seeks to collaborate on studying patient data without disclosing sensitive information. Homomorphic encryption enables the institution to communicate encrypted medical records, compute on encrypted data, and obtain results without ever decrypting the original data. This not only protects patient privacy but also allows for cross-entity collaboration without jeopardizing the confidentiality of critical healthcare data.

5.2.2 Differential Privacy: Injecting Controlled Noise for Anonymity

To safeguard individual privacy during computations or analysis, differential privacy puts a layer of noise into data. The idea is to introduce properly calibrated noise to an algorithm's output, guaranteeing that the inclusion or absence of any single data point has no meaningful impact on the outcome. This strategy is especially useful when the dissemination of statistical data or model forecasts risks exposing personal individual details.

A government entity, for example, that wants to disseminate aggregate statistics regarding the prevalence of particular health issues across different regions could use differentiated privacy. The agency ensures that no individual's health information is identifiable by adding controlled noise into the data before releasing the statistics, ensuring privacy while yet giving significant insights for policymakers and public health planning.

5.2.3 Secure Multiparty Computation: Collaborative Insights Without Data Exposure

Secure multiparty computing (SMPC) enables many entities to cooperatively compute a function over their inputs while keeping those inputs private. This technique is useful in the area of AI when organizations or parties want to collaborate on studies without providing raw data. Each side processes its local data independently, and only the results of the computations are communicated.

Consider a group of financial institutions working together to examine transaction patterns in order to detect probable fraud. Each institution can contribute insights based on its local transaction data through SMPC without revealing specifics about individual transactions. This collaborative approach ensures that the collective intelligence gained from the investigation is solid while protecting sensitive financial information.

5.2.4 Real-World Application: Preserving Privacy in AI-Driven Healthcare Research

A concrete application of these strategies is emerging in the field of AI-driven healthcare research. Consider a coordinated effort among multiple healthcare professionals to construct predictive models for disease outbreaks without providing patient-specific details.

Homomorphic encryption might be used to perform secure computations on encrypted patient records, allowing machine-learning models to be trained without exposing specific health data. Differential privacy ensures that aggregated insights obtained from models are anonymized, preventing unique patients from being identified. Secure multiparty computation enables collaborative model training, allowing healthcare providers to collaboratively contribute to the construction of robust predictive models without divulging the specifics of their patient populations.

This comprehensive approach not only promotes medical research and predictive analytics but also offers a privacy-preserving framework that is consistent with ethical considerations and regulatory obligations in the healthcare sector.

5.2.5 Overcoming Challenges: Trade-Offs and Computational Complexity

While these privacy-protection strategies have numerous advantages, they are not without drawbacks. There are frequently trade-offs between the level of privacy preservation and the utility of the data for research. Injecting noise for differential privacy, for example, may damage precision, necessitating a careful balance to achieve valuable insights while maintaining privacy.

Furthermore, computational complexity presents difficulties, particularly in cases where real-time processing is required. Homomorphic encryption and safe multiparty computation can incur computational cost, requiring optimization efforts to make these techniques practical in resource-constrained situations.

Despite these obstacles, current research and breakthroughs in cryptographic protocols, algorithmic efficiency, and hardware capabilities promise to alleviate these worries, making privacy-preserving AI solutions more practical and scalable.

Navigating the Privacy-AI Landscape

Finally, homomorphic encryption, differential privacy, and safe multiparty computation are critical pillars in the effort to integrate AI into sensitive domains while protecting individual privacy. The complicated dance between these strategies provides a sophisticated approach, allowing enterprises to harness the power of AI without jeopardizing confidentiality.

As artificial intelligence continues to change industries ranging from healthcare to finance, the use of privacy-preserving techniques becomes increasingly important. To strike the correct balance, evaluate the specific use case, the nature of the data involved, and the regulatory landscape. Organizations may leverage the revolutionary promise of AI while adhering to the ethical standards and regulatory obligations that protect individual privacy by navigating the privacy-AI landscape with elegance.

5.3 Implementing Federated Learning for Data Privacy

In the ever-changing environment of artificial intelligence (AI), the need to reconcile innovation with data protection has given rise to innovative solutions. Federated learning is one such disruptive method, a paradigm that changes the locus of machine learning model training away from centralized servers and toward decentralized devices. This section digs into the implementation of federated learning, examining its function in ensuring data privacy while facilitating collaborative model training.

5.3.1 Understanding Federated Learning: Decentralized Intelligence

Federated learning differs from standard machine learning models, which are based on centralized data stores. The model is trained across numerous decentralized devices, such as smartphones, edge devices, or local servers, in a federated learning architecture, eliminating the need to collect raw data at a central server. This decentralized technique takes machine learning to the data source while protecting individual dataset privacy.

Consider the following scenario: a technology company wants to improve its predictive text algorithm for a keyboard application. Rather than gathering and centralizing a wide array of individualized language usage data from users, federated learning allows the model to be trained on the device of each user. The local model updates are then combined to provide insights while never revealing specific user inputs.

5.3.2 Preserving Privacy: Federated Learning in Action

The implementation of federated learning relies around a set of critical processes that contribute to data privacy preservation:

1. Model Initialization: The procedure starts with the creation of a global model. This preliminary model serves as the foundation for training among dispersed devices.

2. Global Model Distribution: The initialized global model is subsequently distributed to all participating devices. Each device has a copy of the model but is unaware of the data of other devices.

3. Local Model Training: The global model is trained using local data on each device. Without transferring raw data to a central server, the device computes model updates based on its own dataset.

4. Model Update Aggregation: Model updates from all participating devices are aggregated, typically using a secure and privacy-preserving technique. This aggregation yields an updated global model that encapsulates the knowledge gained from collective device training.

5. Iterative Refinement: The procedure is performed iteratively, with the revised global model redistributed among devices for additional training. This iterative improvement improves the accuracy of the global model without jeopardizing individual privacy.

5.3.3 Advantages of Federated Learning for Data Privacy

The implementation of federated learning yields several distinct advantages in terms of data privacy:

1. Localized Model Training: Federated learning ensures that model training occurs locally on individual devices. Raw data never leaves the device, mitigating the risk of centralized data breaches or unauthorized access.

2. Minimal Data Movement: Unlike traditional approaches where raw data is centralized for model training, federated learning minimizes the movement of data. Only model updates traverse the network, reducing the potential exposure of sensitive information.

3. User Control: Individuals retain control over their data since training occurs on their devices. This user-centric approach aligns with principles of consent and autonomy, empowering users to actively participate in AI model improvements without compromising privacy.

4. Collaboration Without Centralization: Federated learning facilitates collaborative model training across decentralized devices without the need for a centralized server. This decentralized approach supports collaborative efforts while preserving the confidentiality of individual datasets.

5.3.4 Real-World Applications: From Smartphones to Healthcare

Federated learning is used in a variety of real-world contexts, demonstrating its versatility and efficacy in safeguarding data privacy:

1. Tailored Recommendations: By using federated learning for recommendation systems such as personalized content suggestions or product recommendations, tech organizations can modify models based on user interactions without centralizing sensitive user behavior data.

2. Healthcare Research: In the healthcare arena, federated learning allows for collaborative research on medical data without the need for patient records to be centralized. While protecting individual patient privacy, multiple healthcare organizations can work together to enhance diagnostic models or study treatment outcomes.

3. IoT Predictive Maintenance: Industries that use the Internet of Things (IoT) can use federated learning to do predictive maintenance. In an industrial scenario, edge devices can work together to train models to forecast equipment failures without communicating sensitive operational data to a central server.

4. Financial Fraud Detection: Federated learning can be used to improve fraud detection models in the financial sector. Banks and financial institutions can work together to improve fraud detection algorithms without revealing individual transaction details, safeguarding financial data confidentiality.

5.3.5 Challenges and Considerations

While federated learning appears to be a promising approach to data privacy, it is not without obstacles and considerations:

1. Communication Overhead: Because model updates must be communicated across devices, federated learning creates communication overhead. It is critical to ensure efficient and secure communication lines in order to avoid any bottlenecks.

2. Device Heterogeneity: Devices in a federated learning system may differ in terms of processing capability and network conditions. It is critical for optimal performance to adapt the federated learning process to accommodate this variation.

3. Model Poisoning Attacks: Adversarial entities may attempt to influence model updates in order to jeopardize the global model's integrity. It is critical to implement strong security measures to prevent model poisoning attacks.

4. Federated Averaging Techniques: The aggregation of model changes, which is frequently accomplished by federated averaging, necessitates careful study.

To handle variances in device contributions, weighted averaging or selective aggregation techniques may be required.

5.3.6 Future Directions: Advancing Federated Learning for Privacy

The implementation of federated learning is still evolving, with continuing research resolving obstacles and broadening its application:

1. Differential Privacy Integration: Integrating differential privacy approaches into federated learning can give an extra layer of privacy protection during model training.

2. Advances in Secure Aggregation Protocols: Advances in secure aggregation protocols can further improve federated learning security by guaranteeing that model changes are aggregated in a privacy-preserving way.

3. Synergies with Edge Computing: Federated learning synergizes with edge computing paradigms, allowing for localized model training on edge devices. This convergence has the potential to improve privacy protection while also reducing communication overhead.

4. Efforts to Standardize: Standardizing federated learning protocols and frameworks helps to increase interoperability and widespread adoption. Collaborative efforts to build industry standards are critical for federated learning's seamless incorporation into a variety of applications.

Empowering Privacy Through Federated Learning

And at last, federated learning implementation emerges as a critical technique for safeguarding data privacy in the age of AI. Federated learning empowers users to participate to the advancement of AI models while maintaining ownership over their data by decentralizing model training. Real-world applications from a variety of industries demonstrate the adaptability and impact of federated learning in collaborative model training.

As federated learning matures, addressing problems and embracing future developments, it has the potential to transform the landscape of privacy-preserving AI. The marriage of innovation and data privacy has become a strategic necessity, ensuring that the benefits of AI are realized without jeopardizing the fundamental right to individual privacy.

CHAPTER 6

Data Protection and Compliance

6.1 Regulations and Standards (e.g., GDPR, CCPA) Related to Data Privacy and Security

The primary worry in the fast growing environment of artificial intelligence (AI) and data-driven technologies is the preservation of sensitive information and the protection of individual privacy rights. Regulations and standards provide a framework for responsibly managing personal data in AI systems. They establish guidelines for data collection, storage, processing, and sharing across the AI lifecycle. This helps to ensure that AI development and deployment respect individuals' privacy rights while mitigating potential security risks.

This chapter examines the regulatory frameworks and standards that govern data privacy and security, with a focus on key rules such as the General Data Protection Regulation (GDPR) and the California Consumer Privacy Act (CCPA).

6.1.1 GDPR (General Data Protection Regulation): A Global Standard

The European Union (EU) implemented the General Data Protection Regulation (GDPR) in 2018, marking a watershed event in global data protection. GDPR has far-reaching ramifications for enterprises that manage the personal data of EU people, regardless of their location. GDPR was designed to empower individuals and standardize data privacy rules across Europe.

Personal data is broadly defined under GDPR and includes any information belonging to an identified or identifiable natural person. Organizations that process such data are required to follow a set of key principles:

1. Lawfulness, Fairness, and Transparency: Organizations must process personal data in a lawful, fair, and transparent manner.

2. Purpose Limitation: Data must be acquired for specific, explicit, and legitimate objectives and not processed in a way that contradicts those aims.

3. Data Minimization: Organizations must acquire just the data required for the intended purpose and store it for no longer than necessary.

4. Accuracy: Personal data must be accurate, and actions should be made to ensure that incorrect data is corrected or removed as soon as possible.

5. Storage Restrictions: Data should be retained in a format that allows identification for no longer than is required for the purposes for which it is processed.

6. Integrity and Confidentiality: Organizations must take suitable security measures to prevent unauthorized access, disclosure, alteration, and destruction of personal data.

7. Accountability: Organizations are responsible for adhering to GDPR standards and must be able to verify compliance.

GDPR has established a standard for data protection policies for enterprises all around the world. Because of its extraterritorial reach, businesses operating outside the EU must also follow its principles when managing the data of EU persons. Noncompliance can result in large fines, making GDPR a compelling motivator for businesses to pursue comprehensive data protection procedures.

6.1.2 Consumer Privacy Act of California (CCPA): Innovators in Privacy Legislation in the United States

The California Consumer Privacy Act (CCPA) is a pioneering initiative in the United States to improve consumer privacy rights. The CCPA, which was passed in 2018 and went into effect in 2020, gives California people more control over their personal information collected by businesses.

The following are key provisions of the CCPA:

1. Right to Know: Consumers have the right to know what personal information businesses gather, use, share, or sell.

2. Right to Delete: Subject to specific conditions, consumers can request that corporations delete their personal information.

3. Opt-Out Right: Consumers have the right to refuse the sale of their personal information. Businesses must provide a prominent link on their webpage that says "Do Not Sell My Personal Information."

4. Nondiscrimination: Businesses are not permitted to discriminate against consumers who exercise their privacy rights, such as by charging them different fees or providing a lower level of service.

5. Data Breach Liability: The CCPA provides consumers with a private right of action in the event of a data breach, allowing them to seek statutory damages.

While the CCPA is a state-level policy, its impact extends beyond California due to the state's economic importance. Furthermore, the passage of the California Privacy Rights Act (CPRA) in 2020 reinforces and increases privacy safeguards, indicating a sustained trend toward comprehensive data privacy laws in the United States.

6.1.3 Other International Data Protection Regulations and Standards

Aside from GDPR and CCPA, various more legislation and standards contribute to the global data protection tapestry. These are some examples:

1. Health Insurance Portability and Accountability Act (HIPAA): In the United States, HIPAA is concerned with protecting the privacy and security of medical information.

2. Personal Data Protection Act (PDPA): The PDPA, which was enacted in Singapore, governs the acquisition, use, and disclosure of personal data.

3. Australia's Privacy Act: This legislation governs how federal government agencies and some private sector entities handle personal information.

4. Personal Information Protection and Electronic Documents Act (PIPEDA): In Canada, PIPEDA outlines standards for private sector organizations' collection, use, and disclosure of personal information.

5. ISO/IEC 27701: As an addition to the ISO/IEC 27001 standard, ISO/IEC 27701 establishes, implements, maintains, and continuously improves a Privacy Information Management System (PIMS).

6.1.4 Navigating a Complex Environment: Compliance Challenges and Strategies

Navigating the complicated world of data privacy standards is difficult for enterprises, particularly those that operate on a global scale. Among the most significant challenges are

1. Different regulations apply in different jurisdictions, requiring enterprises to understand and comply with a patchwork of legislation.

2. Data Subject Rights: Organizations face operational obstacles in meeting the numerous rights granted to data subjects, such as the ability to view, modify, or delete their data.

3. Data Security Procedures: Implementing strong data security procedures to guard against breaches and unauthorized access is critical, but it may be time-consuming.

4. Vendor Management: Organizations frequently rely on third-party suppliers for a variety of services, and ensuring that these vendors follow data protection requirements complicates the compliance picture.

To solve these issues, companies can implement comprehensive methods such as

1. Privacy by Design: By including privacy issues into the design and development of goods and systems, data protection is made a priority from the start.

2. Cross-Functional Collaboration: Creating cross-functional teams comprised of legal, IT, security, and compliance experts promotes a comprehensive approach to data protection.

3. Continuous Monitoring and Auditing: Monitoring data processing operations on a regular basis, performing audits, and adopting corrective actions all contribute to continuous compliance.

4. Data Mapping and Classification: Understanding the flow of data within an organization, classifying it based on sensitivity, and putting in place appropriate controls are all critical components of compliance.

5. Employee Education: Educating employees on data protection principles and the specific requirements of applicable regulations is critical for fostering a compliance culture.

6.1.5 Future Legislative Trends in Data Protection

The regulatory environment is being shaped by current and future trends in data protection regulation. Among the significant trends are

1. Global Convergence: Efforts to globalize data protection regulations may gain traction, making compliance easier for multinational corporations.

2. Increased Data Subject Rights: Future rules may increase data subject rights, giving individuals more control over their personal information.

3. Stricter Enforcement: Regulatory agencies are likely to be more forceful in enforcing data privacy regulations, charging significant fines for noncompliance.

4. The Emergence of Industry-Specific Legislation: Certain industries, such as banking and healthcare, may see the emergence of industry-specific data protection regulations targeted to those businesses' unique difficulties.

Navigating the Regulatory Landscape

Lastly, data protection and compliance are the foundations of responsible AI and data-driven innovation. Regulations and standards like GDPR and CCPA raise the bar for enterprises all around the world, requiring them to prioritize privacy and security. Organizations must remain attentive as the regulatory landscape evolves, changing their procedures to suit emerging regulations and ensuring that data protection becomes an inherent part of their operational culture. Navigating this complicated regulatory structure not only protects individual privacy rights but also builds trust between businesses and the people they serve.

6.2 AI's Role in Achieving and Maintaining Compliance

Artificial intelligence (AI) emerges as a powerful ally for enterprises negotiating the complicated environment of data protection in the nuanced dance between technology innovation and regulatory compliance. This chapter dives into the critical role of artificial intelligence (AI) in establishing and maintaining compliance with data protection standards, providing insights into how intelligent technologies may strengthen privacy protections and fortify enterprises against regulatory obstacles.

6.2.1 Recognizing the Impact of AI on Compliance

As enterprises cope with the ever-increasing volume and complexity of data, AI emerges as a transformational force in the compliance domain. The multidimensional nature of artificial intelligence's contribution to compliance can be divided into three important dimensions:

1. Automated Data Mapping and Classification: Artificial intelligence-powered systems excel at sifting through massive datasets to detect, classify, and map data flows inside an organization. This capacity is especially useful for complying with requirements that require a thorough understanding of data processing processes, such as the General Data Protection Regulation (GDPR).

2. Improved Security with Predictive Analytics: AI's predictive analytics capabilities enable enterprises to anticipate future security threats and proactively

apply risk-mitigation measures. AI contributes to the integrity and confidentiality of sensitive information by discovering vulnerabilities and abnormal activity, which is a vital component of complying with data protection standards.

3. Risk Management and Privacy Influence Assessments: AI algorithms can assess the impact of data processing operations on individual privacy, assisting enterprises in conducting privacy impact assessments (PIAs). These assessments are critical to compliance efforts because they provide a systematic method for evaluating and mitigating the risks associated with data processing.

4. Continuous Monitoring and Auditing: AI-driven monitoring solutions allow for real-time monitoring of data processing processes. Organizations can quickly identify deviations from compliance requirements and adopt remedial steps by automating the auditing process. This constant vigilance corresponds to the evolving nature of data protection legislation.

5. Adaptive Compliance Strategies: The ability of AI to learn and adapt positions it as a dynamic compliance tool. Machine learning algorithms can evolve in tandem with regulatory changes, allowing businesses to quickly adapt their compliance strategy in reaction to new legislation or amendments.

6.2.2 Artificial Intelligence in Automated Data Governance

Establishing strong data governance processes is a key aspect of achieving and maintaining compliance. AI is crucial in automating and improving data governance in a variety of ways:

1. Material Discovery and Classification: AI algorithms can automatically locate and classify sensitive data inside large datasets within an enterprise. This is especially useful for firms that must identify and secure personal or sensitive information in order to comply with rules.

2. Policy Enforceability: AI-driven solutions make it easier to implement data governance regulations. Organizations can create data access, sharing, and retention policies, and AI systems can actively monitor and enforce these policies to ensure that data usage fits with compliance requirements.

3. Consent Management: Obtaining and managing user consent for data processing is frequently required in order to comply with rules such as GDPR. AI-driven technologies can help to streamline the consent management process by automating the collecting, tracking, and validation of user consent, ensuring that companies stay within the legal boundaries.

4. Data Quality Management: AI's data profiling and cleansing skills help to maintain data quality, which is an important part of compliance. Data accuracy and dependability are critical for achieving legal requirements and developing trust with data subjects.

5. Incident Response and Breach Detection: When a data breach occurs, AI-driven solutions improve incident response capabilities. These systems can detect anomalies quickly, identify possible security events, and automate responses, limiting the effect of breaches while still adhering to legal demands for timely reporting.

6.2.3 Artificial Intelligence-Driven Privacy Impact Assessments (PIAs)

PIAs are systematic assessments of the potential impact of data processing operations on person privacy. AI is crucial in automating and streamlining the PIA process:

1. Automated Data Flow Analysis: Artificial intelligence systems excel in analyzing and mapping data flows throughout an organization. Automated data flow analysis simplifies the identification of how data goes across various systems, allowing for a more comprehensive knowledge of the privacy implications of data processing.

2. Risk Identification and Mitigation: Using AI's predictive analytics capabilities, possible privacy issues linked with data processing can be identified. AI systems can forecast and analyze the risk of privacy-related incidents by evaluating past data and patterns, allowing enterprises to proactively deploy mitigation measures.

3. Dynamic Compliance Monitoring: AI-driven PIAs are dynamic assessments, not static assessments. As data processing activities change or new regulations are implemented, AI can adapt the PIA process to ensure that organizations continuously evaluate and address privacy risks in real time.

4. Collaboration and Reporting: AI facilitates collaboration among the various stakeholders involved in the PIA process. Automated reporting technologies produce detailed insights and documentation, assisting firms in showing regulatory compliance and fostering transparency.

6.2.4 Addressing Bias and Ethical Issues

While AI adds tremendous value to compliance efforts, it also poses ethical concerns, notably with algorithm prejudice. Biases in AI models can lead to biased outcomes, putting fairness and equity—key objectives of data protection regulations—at risk.

1. Bias Detection and Mitigation: AI tools that detect and mitigate bias contribute to ethical compliance. These tools examine model outputs for biases and make recommendations for changing algorithms to promote fair and unbiased decision-making.

2. Explainability and Openness: Because certain AI models are "black boxes," efforts toward explainability and transparency are critical. Organizations must be able to define how AI-driven choices are made in order to meet data protection rules' openness requirements.

3. Ethical AI Frameworks: By incorporating ethical AI frameworks into compliance strategies, AI applications are ensured to comply to ethical norms. These frameworks guide the development, deployment, and usage of artificial intelligence while complying with the broader ethical considerations inherent in data protection rules.

6.2.5 Obstacles and Considerations

While artificial intelligence has enormous potential for increasing compliance initiatives, numerous obstacles and considerations must be addressed:

1. Interpretability and Explainability: It is a constant effort to ensure that AI-driven judgments are interpretable and explainable. To meet legal standards for openness and accountability, efforts must be made to improve the interpretability of complex AI models.

2. Resource Intensiveness: Implementing AI for compliance can be time-consuming and costly. Organizations may have difficulties due to the price of obtaining and maintaining AI technologies, as well as the requirement for experienced individuals to manage and understand AI systems.

3. Dynamic Regulatory Environment: The regulatory landscape is ever-changing, with frequent modifications and new legislation. Organizations that use AI must be nimble in order to adjust their systems and procedures in response to changing compliance requirements.

4. Data Security Issues: The data on which AI is based for analysis and decision-making is frequently sensitive and subject to data security restrictions. It is critical for compliance to ensure the security of this data, particularly in the context of AI applications.

6.2.6 Future Trends: Evolution of AI-Driven Compliance

Future themes in AI-driven compliance are likely to alter the landscape as AI evolves:

1. AI-Regulatory Convergence: There may be a greater convergence of AI and regulatory frameworks, with rules adopting AI-specific criteria and guidelines.

2. Federated Learning for Privacy-Preserving AI: Federated learning, an AI technique that allows model training across decentralized devices, may gain traction for privacy-preserving AI applications that adhere to data protection standards.

3. AI Governance Frameworks: Organizations can create and implement AI governance frameworks that are compliant with regulatory standards. Within the confines of data protection legislation, these guidelines will guide the ethical and responsible use of AI technologies.

4. Enhanced Collaboration: The level of collaboration between regulatory organizations, industry stakeholders, and AI developers is expected to rise. Collaborations of this type strive to set standards, discuss best practices, and create a collaborative approach to ethical AI research.

A Synergistic Future of AI and Compliance

Therefore, the incorporation of AI into compliance operations heralds a synergistic future in which technology innovation and regulatory conformance coexist. Organizations that use artificial intelligence (AI) to automate data governance, undertake privacy impact assessments, and improve security measures pave the path for a strong and dynamic compliance framework. Navigating the obstacles and ethical concerns associated with AI necessitates a planned and proactive approach to ensure that the potential of intelligent technologies is ethically exploited within the confines of data protection rules. The growing AI-compliance collaboration is a transformative path toward a future in which innovation and regulatory conformance coexist peacefully, fostering trust, openness, and ethical data practices.

CHAPTER 7

Securing AI Models

7.1 Best Practices for Securing AI Models

As enterprises increasingly rely on AI models to derive insights and power important decision-making processes, the need to secure these models becomes critical. This chapter investigates best practices for protecting AI models, including a complete methodology that tackles vulnerabilities, protects sensitive data, and ensures the ethical deployment of intelligent systems.

7.1.1 Strong Data Governance as the Basis

Before getting into specific AI model security measures, it's critical to establish strong data governance policies. Data governance serves as the foundation for AI models, influencing the quality, integrity, and privacy of the data they handle. The following are important data governance considerations:

1. Data Quality Management: Ensure that data is accurate, full, and reliable. To resolve inaccuracies and inconsistencies, implement data quality checks and cleansing methods.

R. R. Palle and K. C. R. Kathala, *Privacy in the Age of Innovation*,
https://doi.org/10.1007/979-8-8688-0461-8_7

2. Privacy by Design: Integrate privacy concerns into the design and deployment of AI algorithms. To protect individual privacy, conduct privacy impact assessments and follow data protection legislation.

3. Access Controls: Use strong access controls to limit data access based on roles and responsibilities. Only authorized people should have access to sensitive datasets and model training environments.

4. Transparency: Encourage openness in data processing processes. To develop trust with users and stakeholders, clearly express data usage restrictions, model outputs, and the reasoning behind AI-driven decisions.

7.1.2 Model Development and Security Training

Vigilance is needed throughout the development process to build secure AI models. This entails examining the training data in addition to protecting the model itself. In order to reduce the possibility of biases or vulnerabilities being included into the model, one should be aware of the sources of the trained data and their security protocols.

AI model security begins throughout the development and training phases. Implementing strong security measures throughout these stages protects against potential vulnerabilities:

1. Model Poisoning Attacks: Artificial intelligence models are vulnerable to adversarial attacks, in which malevolent actors modify input data to deceive the model. To increase resilience against adversarial threats, use approaches such as adversarial training and input validation.

Prevent model poisoning attacks by verifying and cleaning training data. During model training, use anomaly detection techniques to identify malicious inputs.

2. Secure Model Repositories: Prevent unauthorized access to model repositories. Encryption and access controls should be implemented to ensure that only authorized users can install or change models.

3. Version Control and Auditing: Keep AI models under version control to track changes and updates. Implement auditing measures to keep track of model development efforts and detect unauthorized changes.

7.1.3 Operational Security and Deployment

Once AI models are ready for deployment, a strong operational security architecture is required to assure long-term security:

1. Protect APIs: If the AI model is accessed via APIs, protect these interfaces with robust authentication and encryption. Monitor API usage on a regular basis to recognize and respond to unusual behaviors.

2. Data Encryption: Encrypt data in transit as well as at rest. Encryption isn't just used to protect operational AI models. To ensure complete security, data should be encrypted throughout its lifecycle. This includes encrypting data at rest (storage), in transit (transfer), and even in use (AI model processing). This layered approach reduces the risk of unauthorized access or

data breaches at all points. Encryption procedures should be used to protect data during the model's operating phase, especially when dealing with sensitive information.

3. Continuous Monitoring: Monitor model performance and behavior in real time. Set up warnings for unexpected patterns or deviations that could indicate a security breach or model drift.

4. Container Security: To prevent exploitation, secure containers using best practices such as least privilege access, regular upgrades, and vulnerability assessment when employing containerized environments.

7.1.4 Ethical Concerns and Bias Reduction

Beyond the technological issues, ethical considerations and bias reduction are critical to responsibly protecting AI models. This section has already been discussed in Chapter 6. You can read that chapter to understand this concept.

7.1.5 Education and Awareness of Users

Security is a shared responsibility, and user education is critical to maintaining a secure AI environment:

1. Training and Awareness Programs: Hold frequent training sessions to educate users on best practices for security. Raising awareness of potential dangers, phishing efforts, and the significance of following security policies.

2. Secure Development Techniques: Educate developers on secure coding practices, emphasizing the necessity of designing secure code and performing thorough security audits throughout the development lifecycle.

3. Incident Response Training: Provide staff with the skills and information required to effectively respond to security issues. Simulated drills should be carried out to ensure a quick and coordinated reaction in the event of a security breach.

4. User Access Reviews: Review and adjust user access privileges on a regular basis. Reduce the risk of unauthorized access by ensuring that people have the proper permissions for their jobs and responsibilities.

7.1.6 Adherence to Regulatory Standards

Compliance with regulations is critical for companies that handle sensitive data. Security measures must comply with regulations that apply to a variety of stakeholders, including enterprises, data providers who contribute training data, AI model developers, and even government agencies. This ensures a comprehensive approach to data security throughout the AI ecosystem. Important considerations include

1. Security Controls and Regulations: Identify and connect important regulatory standards, such as GDPR, HIPAA, or industry-specific requirements, to security controls. This guarantees that security measures are in line with regulatory requirements.

2. Periodic Compliance Audits: Conduct regular audits
    to ensure that regulatory criteria are being met. This
    includes assessing the efficiency of security controls,
    data protection safeguards, and compliance with
    privacy regulations.

3. Data Subject Rights: Put in place measures to make
    it easier for data subjects to exercise their legal
    rights. This contains procedures for requesting data
    access, data portability, and the right to be forgotten.

4. Reporting and Documentation: Maintain detailed
    documentation of security and compliance
    initiatives. Prepare detailed reports for regulatory
    agencies to demonstrate your commitment to data
    protection and security.

7.1.7 Information Sharing and Collaboration

Security is a constantly changing landscape, and collaboration is essential
for staying ahead of developing threats:

1. Platforms for Information Sharing: Participate in
    industry-specific information sharing platforms to
    stay up to date on the most recent security risks and
    vulnerabilities. Sharing insights and best practices
    for risk mitigation helps the collective defense
    against evolving threats.

2. Interaction with Peers: Encourage industry
    colleagues to collaborate. Participate in forums,
    working groups, and collaborative projects to
    exchange ideas and experiences around securing
    AI models.

3. Integration of Threat Intelligence: Integrate threat intelligence feeds into security operations. Using real-time threat intelligence improves the ability to detect and respond to emerging threats in a proactive manner.

4. Establish Routes for Responsible Vulnerability Disclosure: Create channels for responsible vulnerability disclosure. Encourage ethical hackers and security researchers to report vulnerabilities, facilitating prompt mitigation.

7.1.8 Incident Response and Recovery Planning

Despite robust security measures, incidents may still occur. A well-defined incident response and recovery plan is crucial:

1. Incident Response Team: Form an incident response team comprising individuals with expertise in cybersecurity, legal, and communication. This team should be ready to activate in the event of a security incident.

2. Incident Detection and Analysis: Implement tools and processes for timely detection and analysis of security incidents. Leverage AI-driven tools to augment human capabilities in identifying and understanding potential threats.

3. Communication Protocols: Define communication protocols for internal and external stakeholders in the event of a security incident. Transparent and timely communication helps manage the impact on users and organizational reputation.

4. Post-Incident Analysis: Conduct thorough post-incident analysis to understand the root causes of security incidents. Use insights gained to refine security measures and enhance resilience against similar incidents in the future.

7.1.9 Emerging Technologies and Adaptive Security

As technology evolves, security measures must also adapt to emerging threats. Considerations for adapting to new technologies include

1. Quantum-Safe Encryption: Anticipate the impact of quantum computing on encryption. Explore quantum-safe encryption techniques to ensure the long-term security of sensitive data.

2. AI in Threat Detection: Leverage AI in threat detection mechanisms. AI-driven threat detection systems can analyze vast datasets and identify patterns indicative of emerging threats, enhancing the organization's ability to proactively respond.

3. Blockchain for Data Integrity: Explore the use of blockchain technology to ensure the integrity and immutability of critical datasets. Blockchain can enhance transparency and trust in data transactions.

4. Adaptive Security Models: Move toward adaptive security models that continuously assess and adjust security measures based on evolving threats. Dynamic and adaptive security frameworks enhance resilience against sophisticated attacks.

7.1.10 Regular Security Assessments and Reviews

Continuous improvement is key to maintaining the effectiveness of security measures. Regular assessments and reviews contribute to ongoing refinement:

1. Penetration Testing: Conduct regular penetration testing to identify vulnerabilities in AI models and associated infrastructure. Addressing identified vulnerabilities strengthens the overall security posture.

2. Security Audits: Engage in regular security audits conducted by internal or external teams. Audits assess compliance with security policies, evaluate the effectiveness of security controls, and identify areas for improvement.

3. Model Performance Monitoring: Implement ongoing monitoring of AI model performance, including its security aspects. Regularly review model outputs, assess for potential anomalies or deviations, and adjust security measures accordingly.

4. Stakeholder Feedback: Solicit feedback from users and stakeholders regarding the security of AI models. User feedback can provide valuable insights into potential security concerns and areas for enhancement.

A Holistic Approach to AI Model Security

Securing AI models is a multidimensional task that necessitates a comprehensive approach that includes strong data governance, ethical concerns, and proactive security measures. Organizations may bolster their defenses against new threats, preserve sensitive data, and ensure the responsible deployment of intelligent systems by incorporating best practices across the AI development lifecycle. As the technology and security landscapes advance, the industry's dedication to continuous improvement and collaboration will be critical in sustaining the resilience and trustworthiness of AI models.

7.2 Model Explainability and Fairness

Two essential elements deserve special attention in the complex terrain of safeguarding AI models: model explainability and fairness. This chapter digs into the significance of these factors in the context of AI security, examining how transparency and equity contribute not just to ethical AI deployment but also to effective security measures.

7.2.1 The Importance of Model Explicability

Model explainability, also known as interpretability, is the degree to which humans can understand an AI model's internal mechanisms and decision-making processes. This transparency is critical for a number of reasons:

1. Accountability and trust: Understanding how an AI model makes a decision encourages trust among users, stakeholders, and the larger community. Transparent models improve accountability by enabling organizations to explain and justify AI-driven actions.

2. Ethical Considerations: Transparency is required for ethical AI deployment. Users have the right to know how their data is being used and how artificial intelligence judgments may affect their lives. Model explainability is consistent with ethical norms, ensuring that people are aware of the reasons impacting AI-driven results.

3. Detection and Mitigation of Biases: Transparent models make it easier to detect and mitigate biases. Organizations can identify and resolve potential biases in AI systems by offering insights into the features and patterns driving predictions.

4. Regulatory Compliance: Transparency in automated decision-making processes is required by several data protection and privacy rules. Model explainability assists firms in meeting regulatory standards, such as the General Data Protection Regulation's right to explanation (GDPR).

7.2.2 Model Explainability Techniques

Several strategies can improve AI model explainability by providing insights into their decision-making processes:

1. Feature Importance Analysis: By analyzing the importance of input features, users can learn which elements have a substantial influence on the model's predictions. SHAP (SHapley Additive exPlanations) values, for example, give a quantitative assessment of feature importance.

2. LIME (Local Interpretable Model-agnostic Explanations): LIME generates locally faithful explanations for particular predictions, revealing how specific cases influenced the model's conclusion. This method is very beneficial for complex models.

3. Decision Trees and Rule-Based Models: Decision trees and rule-based models are transparent in their decision-making processes by definition. These models are simpler to interpret, making them appropriate for applications where explainability is important.

4. Layer-wise Relevance Propagation (LRP): LRP assigns relevance scores to each input feature, showing how the feature affects the model's output. This approach is useful for neural networks and deep learning models, as it provides information about feature contributions.

7.2.3 The Fairness Imperative in AI Models

Fairness in AI models is both an ethical and societal requirement. Fairness entails treating all persons and groups equally, regardless of demographic features. Addressing fairness in AI is critical for several reasons:

1. Avoiding Discrimination: AI model biases might result in discriminatory outcomes, exacerbating existing socioeconomic inequities. Fair AI policies seek to reduce bias and guarantee that AI systems do not disproportionately affect specific populations.

2. User Adoption and Trust: Fair AI models
    increase user trust and adoption. Users are more
    inclined to engage with an AI system and trust its
    recommendations when they believe it treats them
    fairly, promoting great user experiences.

3. Compliance with Legal and Regulatory
    Requirements: Discriminatory AI practices may
    breach antidiscrimination laws and regulations.
    Ensuring fairness in AI models correlates with
    legal and regulatory standards, helping to ensure
    compliance with frameworks like the Civil Rights
    Act and the Fair Housing Act.

4. Ethical Deployment: Ethical concerns require that
    AI models not reproduce or magnify societal biases.
    Fair AI methods contribute to the responsible and
    ethical deployment of AI by harmonizing with
    justice and equality values.

7.2.4 Difficulties in Ensuring Model Fairness

Achieving fairness in AI models is not without difficulties, and overcoming
these difficulties is critical for responsible AI deployment:

1. Bias in Training Data: AI algorithms can propagate
    biases in training data. If the training data
    represents current societal biases, the model may
    learn and reproduce them. To address this issue,
    careful curation and preprocessing of training data
    are required.

2. Lack of Diversity in Development Teams: The makeup of AI development teams can have an impact on model fairness. Inadequate diversity may result in oversight or unintended bias. Diverse teams contribute a variety of perspectives, which aids in identifying and correcting potential biases during model building.

3. Complex Model Architectures: Complex models, particularly deep learning models, can have opaque decision-making processes. Understanding and eliminating biases in these models can be difficult, necessitating the use of specific fairness assessment approaches and tools.

4. Changing Definitions of Fairness: Fairness is a very subjective and context-dependent concept. Different stakeholders may define fairness differently. Organizations must negotiate the changing environment of fairness definitions and collaborate with stakeholders to develop fairness criteria for specific applications.

7.2.5 Model Fairness Assurance Techniques

To address fairness in AI models, approaches for identifying, measuring, and mitigating biases must be implemented. Among the most important techniques are

1. Markers of Fairness: Use indicators of fairness to analyze differences in model predictions across demographic groupings. Disparate impact and equalized odds metrics can provide quantitative measurements of fairness.

2. Adversarial Training: Include adversarial training to strengthen models against adversarial attacks aimed at exploiting biases. Adversarial training involves purposefully biased data being used to train models in order to improve their robustness.

3. Algorithms for Detecting and Correcting Biases: Use bias detection algorithms to discover and quantify biases in model predictions. Then, use bias correction strategies to reduce the influence of biases on decision-making outcomes.

4. AI That Can Be Explained: Use explainable AI techniques not only for transparency but also for determining fairness. Explainable AI can help identify potential biases by revealing how specific features contribute to predictions.

7.2.6 Ongoing Monitoring and Bias Reduction

To ensure fairness, implement real-time monitoring of model outputs for potential biases. Set up alerts to detect and address biases as they occur, preventing unfair behaviors from continuing.

1. Iterative Model Training: Use an iterative model training approach. Continuously improve models based on feedback and fairness performance measures. Update training data on a regular basis to address evolving biases and guarantee models remain fair over time.

2. Stakeholder Engagement: Collect feedback on the fairness of AI models from stakeholders such as end users and affected communities. Include stakeholder input in the model development and refining processes.

3. Bias Impact Assessments: Conduct impact assessments to comprehend the real-world implications of model projections. Examine how AI-driven decisions may affect various demographic groups and update models accordingly to prevent negative consequences.

7.2.7 The Relationship Between Model Explainability and Fairness

Model explainability and fairness are characteristics of ethical AI deployment that are interconnected. Transparent models promote justice by enabling stakeholders to understand and examine decision-making processes. Addressing biases and ensuring fairness, on the other hand, increases the ethical implications of AI, increasing the need for explainability.

1. Explainability for Fairness: Transparent models aid in the discovery of biases by revealing how input features influence decisions. This transparency allows stakeholders to analyze if certain groups are disproportionately harmed by the model's projections, which helps to determine fairness.

2. Stakeholder Empowerment: When explainability and fairness intersect, stakeholders are empowered. By exploiting insights gained through transparent

and explainable decision-making, users, affected groups, and regulatory authorities may hold enterprises accountable for the fairness of AI models.

3. Ethical Points to Consider: Transparency in model explanations and fairness in AI applications are consistent with larger ethical concerns. Organizations must negotiate the ethical dimensions of AI deployment by ensuring that their AI initiatives include both model explainability and justice.

7.2.8 The Way Forward: Ethical and Secure AI Models

Finally, safeguarding AI models extends beyond technical measures to include ethical considerations, transparency, and fairness. Model explainability and fairness are not separate notions but rather components of responsible AI deployment. Organizations must take a holistic approach, combining model explainability and fairness methodologies across the AI development lifecycle.

The way forward entails a commitment to ongoing development, collaboration, and stakeholder engagement. As technology advances, so must the tactics for protecting AI models. Organizations can develop a foundation for AI deployment that is not only secure but also aligned with societal values, encouraging trust and accountability in the ever-changing field of artificial intelligence by prioritizing transparency, fairness, and ethical considerations.

7.3 Model Deployment Security

Model deployment security appears as a critical part of assuring the integrity, confidentiality, and availability of AI systems in the dynamic context of artificial intelligence (AI). This chapter delves into the difficulties, best practices, and emerging trends that businesses must consider to defend their AI deployments against growing threats, diving into the many facets of model deployment security.

7.3.1 The Value of Model Deployment Security

Model deployment is an important stage in the AI lifecycle since it signals the move from development to real-world application. Security considerations are critical at this level for various reasons:

1. Sensitive Data Protection: Deployed AI models frequently work on sensitive data, ranging from personal information to private company data. It is critical to secure the deployment environment in order to prevent illegal access and protect this vital information.

2. Adversarial Threat Mitigation: As AI systems become more prevalent, they become increasingly appealing targets for adversarial assaults. Model deployment security is critical for minimizing hostile threats that seek to manipulate or exploit flaws in deployed models.

3. Continuous Monitoring and Compliance: AI systems can be continuously monitored thanks to security measures implemented during model deployment. This is critical for meeting data protection legislation, industry standards, and internal security rules.

4. User Confidence and Trust: Security failures diminish trust in AI systems. Organizations develop and retain user confidence by emphasizing model deployment security, hence strengthening the dependability and trustworthiness of AI-driven solutions.

7.3.2 Model Deployment Security Issues

Securing the deployment of AI models involves specific difficulties that organizations must overcome in order to maintain a strong security posture:

1. Complex Infrastructure: Model deployment frequently incorporates complex infrastructures such as cloud services, edge computing, and containerized environments. Securing these many and interrelated components necessitates comprehensive strategies that take into account the whole deployment pipeline.

2. Concerns About Data Privacy: Models that have been deployed interact with sensitive data, creating worries about data privacy. To protect user data during model inference, organizations must incorporate encryption, access limits, and other privacy-preserving methods.

3. Changing Threat Environment: The threat landscape is always changing as enemies devise new methods. Organizations require adaptive security solutions that can respond to emerging threats while also assuring the long-term durability of deployed AI models.

4. Integration with Existing Systems: Integrating AI models with existing IT infrastructure can be difficult, particularly in large businesses with complex legacy systems. Maintaining security while ensuring smooth integration necessitates careful planning and implementation.

7.3.3 Model Deployment Security Best Practices

Adhering to recommended practices is critical for firms seeking to improve the security of AI model deployment. Important considerations include

1. Model Packaging and Delivery Security

   – Container Security: Prioritize container security while employing containerized environments. To mitigate vulnerabilities, use recommended practices such as image scanning, least privilege access, and regular updates.

   – Code Signing: Code signing is used to validate the integrity and validity of model packages. This ensures that only approved and unaltered models are used.

2. Authentication and Access Controls

   – Role-Based Access Control (RBAC): Use RBAC to limit access to deployed models based on their roles and responsibilities. To avoid illegal model updates, restrict access to authorized people.

   – Mechanisms of Authentication: Strong authentication measures should be used to authenticate the identity of users interacting with deployed models. Multifactor authentication is included as an extra layer of security.

3. Data Encryption in Transit and at Rest

   – Transport Layer Security (TLS): Enable TLS for secure communication between deployment infrastructure components. This assures the security and integrity of data sent during model inference.

   – Data Encryption: Encrypt data at rest, particularly when storing model parameters, configurations, or other sensitive information. Encryption protects stored data from unauthorized access.

4. Anomaly Detection and Continuous Monitoring

   – Real-Time Monitoring: Monitor model performance and interactions in real time. Set up alerts for unusual behavior that could indicate a security incident or a divergence from typical norms.

   – Anomaly Detection: Use anomaly detection techniques to spot out-of-the-ordinary phenomena in model inputs or outputs. This improves the detection of adversarial assaults and unusual activity.

5. Software Updates and Patch Management on a Regular Basis

   – Patch Vulnerabilities: Update software dependencies, libraries, and frameworks used in the deployment environment on a regular basis. Patching known vulnerabilities on time reduces the likelihood of adversary exploitation.

   – Version Control: Keep track of the versions of deployed models and associated software components. This makes it easier to trace changes, roll back to prior versions if necessary, and ensure the consistency of deployed environments.

6. Incident Response Planning

   – Incident Response Team: Create an incident response team with AI security experience. This team should be prepared to respond quickly to security issues, conduct investigations, and put corrective measures in place.

   – Protocols for Communication: In the case of a security incident, establish explicit communication channels for notifying stakeholders and users. Transparent communication aids in managing the impact and preserving trust.

7. Encrypted APIs and Endpoints

   – API Security: If models are accessed via APIs, these interfaces should be secured with strong authentication techniques, API keys, and rate limits. To avoid API vulnerabilities, use secure coding methods.

   – Endpoint Security: Ensure endpoint security where AI models are installed. Implement security mechanisms at the application layer to protect against typical vulnerabilities such as injection attacks.

8. Observance of Regulatory Standards

   – Regulatory Mapping: Align security measures with applicable regulatory standards like GDPR, HIPAA, or industry-specific criteria. Audit and assess compliance with these standards on a regular basis to meet legal and regulatory requirements.

- Data Subject Rights: Establish procedures for dealing with requests for data subject rights, such as access, rectification, and deletion. Complying with these requests helps to increase transparency and accountability.

9. DevOps Security Practices

- Integration of DevSecOps: Incorporate security practices throughout the DevOps lifecycle. Adopt DevSecOps principles to guarantee that security is woven throughout all stages of model creation, deployment, and operation.

- Automated Security Testing: Include automated security testing in the pipeline for continuous integration and continuous deployment (CI/CD). Automated testing aids in the detection and remediation of security flaws early in the development process.

10. Employee Training and Awareness

- Security Training: Conduct regular security training sessions for staff participating in the deployment of AI models. Increase understanding of security best practices, potential dangers, and the importance of following security rules.

- Phishing Awareness: Inform employees about phishing dangers and best practices for detecting and avoiding phishing attacks. Phishing is still a common source of security breaches and requires aggressive prevention.

7.3.4 Emerging Model Deployment Security Trends

As technology advances, new trends and developments shape the model deployment security landscape:

1. Integrating Explainable AI Techniques in Security Audits: Integrating explainable AI techniques in security audits improves the transparency of security measures. Explainability helps auditors and stakeholders understand how security protocols work, resulting in a more complete and informed evaluation.

2. AI Zero Trust Architecture: In AI deployment, the Zero Trust security model, which implies that no entity, internal or external, can be trusted, is gaining traction. By requiring verification and authorization for every access attempt, Zero Trust principles improve security.

3. Secure Federated Learning: Organizations are investigating federated learning as a safe method of modeling training across decentralized contexts. This technique enables models to be trained cooperatively without the need for raw data exchange, hence resolving privacy concerns associated with centralized training.

4. AI-Driven Security Operations Integration: AI is increasingly being incorporated into security operations for real-time threat identification and response. Machine learning algorithms are used in AI-driven security operations to evaluate patterns, detect anomalies, and automate responses to security issues.

5. Model Integrity via Blockchain: The use of blockchain technology to secure the integrity of deployed models is gaining pace. Blockchain provides an immutable and transparent ledger that prevents unauthorized model updates and ensures model authenticity.

7.3.5 Future Model Deployment Security Considerations

In the future, companies must be watchful and proactive in addressing developing model deployment security challenges:

1. Quantum-Safe Encryption: The advent of quantum computing puts standard encryption approaches in jeopardy. Exploring and implementing quantum-safe encryption solutions will be critical to ensuring model deployment security in the future.

2. AI-Driven Threat Hunting: Using AI for proactive threat hunting and identifying possible weaknesses in AI models will become standard practice. AI-driven threat hunting improves the detection of sophisticated threats early in the deployment lifecycle.

3. International Collaboration on AI Security Standards: As AI becomes a worldwide phenomenon, international standards for AI security must be developed. Collaboration across governments, companies, and regulatory organizations will help to create a uniform strategy to AI model deployment security.

4. Dynamic Security Policies: It will be critical to have dynamic security policies that respond to evolving threats and compliance requirements. To handle emerging risks in the quickly changing AI ecosystem, organizations require flexible and responsive security frameworks.

Protecting AI Model Deployment's Future

To summarize, model deployment security is a critical component of responsible and effective AI implementation. Organizations can strengthen their AI installations against a variety of dangers by implementing best practices, maintaining current emerging trends, and planning for future issues. Integrating security into all phases of the AI lifecycle, from development to deployment and operations, enables a holistic and robust strategy to protecting the future of AI model deployment. As technology advances, a commitment to proactive security measures will be critical in establishing a trustworthy and secure AI ecosystem.

CHAPTER 8

Case Studies

8.1 Real-World Examples of AI Enhancing Information Security and Data Privacy

The incorporation of artificial intelligence (AI) has shown to be a revolutionary force in the ever-changing world of information security and data privacy. AI is being used by businesses in a variety of industries to improve cybersecurity and protect sensitive data. This chapter dives into real-world examples that demonstrate AI's practical impact on information security and data privacy.

8.1.1 Healthcare Industry: Threat Prevention Using Predictive Analytics

The necessity of protecting patient data in the healthcare industry cannot be emphasized. To identify and mitigate potential security issues, hospitals and healthcare institutions are increasingly turning to AI-driven predictive analytics. AI systems can identify and mitigate future breaches by evaluating trends and abnormalities in data access. This proactive strategy not only safeguards sensitive patient data but also ensures the integrity of vital healthcare systems.

8.1.2 Financial Institutions: Detection and Prevention of Fraud

Cybercriminals seeking to exploit flaws in financial institutions' systems pose a persistent danger. Artificial intelligence has emerged as a critical ally in the fight against fraud. Machine learning algorithms that have been trained on massive datasets of historical transactions can quickly discover anomalous patterns that indicate fraudulent conduct. This real-time detection capability enables financial institutions to respond quickly, preventing unwanted access and securing their clients' financial data.

8.1.3 E-commerce: Behavioral Analysis for Personalized Security

E-commerce platforms manage massive volumes of client data, including personal and financial data. Artificial intelligence (AI) technologies, particularly behavioral analysis, are being used to develop personalized security measures. AI systems can detect abnormalities that may suggest account compromise by analyzing and learning from user behavior. This not only improves the overall security of the e-commerce ecosystem, but it also gives users a more smooth and personalized online shopping experience.

8.1.4 Manufacturing: Industrial Internet of Things Security

Manufacturing processes have become more interconnected as the Industrial Internet of Things (IIoT) has grown in popularity, resulting in a larger attack surface for cyberattacks. By continuously monitoring network traffic and identifying potential vulnerabilities, AI plays a critical role in securing industrial environments. Smart artificial intelligence systems

106

can detect aberrant activity, such as unwanted access attempts or strange data patterns, ensuring the integrity and security of vital manufacturing processes.

8.1.5 Government: Critical Infrastructure Protection

Government entities in charge of key infrastructure like power grids and transportation networks are prominent targets for cyberattacks. AI is being used to improve the security of these critical systems. AI systems can examine vast information to discover potential risks and weaknesses using advanced threat detection algorithms and machine learning. This proactive approach protects national security by preventing disruptions to key services.

8.1.6 Social Media Platforms: Moderation of Content and User Privacy

Because social media platforms handle massive amounts of user-generated content, they are vulnerable to data breaches and the propagation of dangerous content. Content moderation technologies driven by AI analyze text, photos, and videos in real time to detect and delete unsuitable or harmful content. Furthermore, AI algorithms can improve user privacy by incorporating features like facial recognition opt-out and customizable privacy settings.

8.1.7 Education Sector: Student Data Security

The education sector provides a unique environment for AI applications, particularly in the field of EdTech. Intelligent tutoring systems (ITS) are a prominent example. These AI-powered systems tailor learning experiences by analyzing student data such as performance metrics, learning styles, and progress. While ITS provides numerous advantages, it also raises specific data privacy concerns.

Security measures must address potential risks such as unauthorized access to student data, data breaches, or the use of student information for profiling or targeted advertising. The collection, use, and disclosure of student data is governed by regulations such as the Family Educational Rights and Privacy Act (FERPA) in the United States, which require educational institutions to implement appropriate safeguards. EdTech can thrive responsibly by striking a balance between using AI to personalize learning and protecting student privacy.

8.2 Success Stories and Lessons Learned

The use of artificial intelligence (AI) in information security and data privacy has led to multiple success stories across industries. These success examples not only show the practical benefits of AI, but they also give vital insights for enterprises looking to improve their cybersecurity posture. This chapter looks at prominent success stories and distills key lessons learned from them.

8.2.1 Financial Sector: Increasing Fraud Detection Efficiency

To fight the mounting issues of financial fraud, a large worldwide bank developed AI-driven fraud detection tools. The bank realized a considerable reduction in false positives by employing machine learning algorithms, allowing for more accurate identification of suspicious activities. This initiative's success not only resulted in significant cost savings, but it also demonstrated the significance of continuous learning for AI models. The lesson here is that regular AI algorithm updates and refining are critical for staying ahead of emerging dangers.

8.2.2 Technology Firm: Cloud Infrastructure Security

A well-known technology firm successfully used AI to improve the security of its cloud infrastructure. The company spotted and stopped unwanted access attempts in real time using AI-driven anomaly detection. This initiative's success underlined the necessity for a comprehensive strategy to cybersecurity, emphasizing the integration of AI with other security measures. The lesson learned is that AI should be used in conjunction with existing security processes to provide a solid defense against cyber threats.

8.2.3 Healthcare Provider: AI for Patient Data Privacy Protection

A healthcare provider used privacy-preserving artificial intelligence algorithms to evaluate patient data while adhering to tight privacy standards. The supplier was able to obtain significant insights from sensitive health information without jeopardizing individual privacy by utilizing techniques such as homomorphic encryption. This success story underscores the significance of ethical and regulatory compliance in AI implementations. When adopting AI in sensitive fields, enterprises must emphasize privacy and comply with appropriate rules.

8.2.4 Ecommerce Powerhouse: Customized Security Measures

An e-commerce behemoth used AI-driven behavioral analysis to improve user security and avoid account hacks. The company developed tailored security features that adapt to individual usage habits by evaluating user behavior patterns. This initiative's success demonstrated the effectiveness of user-centric security measures. The lesson learned is that AI should focus on understanding and reacting to user behaviors in order to provide a better user-friendly security experience.

8.2.5 Energy Sector: Artificial Intelligence for Predictive Maintenance and Security

In the energy industry, a utility firm used AI for both predictive maintenance and critical infrastructure security. The business forecasted future breakdowns and vulnerabilities by analyzing sensor data from industrial equipment, minimizing downtime and improving overall security. This success story demonstrates how AI may improve operational efficiency while also strengthening cybersecurity. The lesson learned is that AI can provide diverse solutions that address both operational and security concerns at the same time.

Adaptive Security Measures in Educational Institutions

To protect student data and prevent illegal access, an educational institution adopted AI-driven adaptive security measures. Based on contextual criteria such as user location and device information, the system constantly changed security procedures. This initiative's success underlined the significance of agility and adaptation in AI-driven security solutions. The lesson learned is that AI systems must be adaptable to changing circumstances and emerging dangers.

These success stories and lessons learned highlight AI's disruptive impact in information security and data privacy. As enterprises continue to take inspiration from these experiences, they will be able to confidently navigate the developing cybersecurity landscape, embracing AI as a powerful ally in the pursuit of greater security and privacy.

CHAPTER 9

AI in Data Privacy and Ethics

9.1 The Ethical Considerations of AI in Privacy

In an era of rapid technological growth, the incorporation of artificial intelligence (AI) in various facets of our life raises critical ethical concerns, notably in the field of privacy. As organizations use the power of AI to promote innovation, it becomes critical to negotiate the ethical landscape to ensure that privacy remains a fundamental and protected right. This chapter dives into the ethical implications surrounding AI in privacy, examining the challenges and opportunities that occur in this dynamic confluence.

9.1.1 Balancing Innovation and Individual Anonymity Preservation

Finding the delicate balance between innovation and personal information protection is one of the most important ethical considerations in employing AI for privacy-related jobs. As corporations attempt to

© Ranadeep Reddy Palle, Krishna Chaitanya Rao Kathala 2024
R. R. Palle and K. C. R. Kathala, *Privacy in the Age of Innovation*,
https://doi.org/10.1007/979-8-8688-0461-8_9

offer more efficient and personalized AI-driven services, there is a risk of unwittingly invading individual privacy. To strike a balance, AI systems must be designed and developed with privacy in mind. This includes incorporating privacy-by-design principles and ethical issues into the heart of AI systems.

Privacy is inextricably tied to the concept of anonymity, and ethical AI practices necessitate a commitment to preserving individual anonymity whenever possible. To reduce the risk of re-identification, organizations using AI should use privacy-preserving approaches such as anonymization and aggregation. Finding the correct balance between data utility and individual anonymity is a subtle ethical consideration that necessitates a deliberate approach to data processing and distribution.

9.1.2 Informed Consent and Transparency

Respecting individuals' autonomy and obtaining informed permission is a cornerstone of ethical AI in privacy. Users must be sufficiently informed about the data collecting and processing practices of AI systems in order to make informed decisions about sharing their information. Transparency in AI algorithms is critical because it creates trust and allows people to understand how their data is being utilized. Ethical AI methods necessitate clear communication and user-friendly interfaces that enable users to exert control over their privacy preferences.

9.1.3 Avoiding Discrimination and Bias

The ethical implications of AI in privacy extend to the potential reinforcement of existing prejudices or the unintended formation of new ones. AI systems, if not carefully developed and trained, can unwittingly perpetuate discrimination, resulting in unequal effects on some demographic groups. Ethical issues dictate that enterprises actively address prejudice in AI systems to ensure fairness and equity.

Implementing diverse and representative datasets and periodically assessing AI models for bias are critical steps toward fostering ethical AI in privacy.

9.1.4 Reducing Invasion

The ethical use of AI in privacy demands enterprises to reduce the intrusiveness of their systems. This entails assessing the granularity of data collection and processing, ensuring that only necessary and relevant information is accessed. Ethical considerations necessitate that AI systems should not participate in unwarranted monitoring or data collection that may infringe on individuals' private rights. Maintaining ethical standards requires striking a compromise between the utility of AI applications and minimizing intrusiveness.

9.1.5 Accountability and Responsibility

Ethical AI practices in privacy demand a clear framework for accountability and responsibility. Organizations must accept responsibility for the implications of their AI systems, including any privacy violations or unforeseen repercussions. Establishing strong governance structures, conducting regular audits, and cultivating an accountability culture are critical components of ethical AI implementation. Ethical issues also extend to enterprises' responsibilities to mitigate any negative consequences and continuously enhance their AI systems in response to developing ethical challenges.

9.1.6 Global Standards and Compliance

Navigating the ethical implications of AI in privacy gets considerably more complicated in a global context with varied rules and cultural standards. Ethical AI methods necessitate firms staying current on international privacy requirements such as the General Data Protection Regulation (GDPR) and other regional rules. Compliance with these standards not only assures legal conformance but also coincides with ethical ideals, demonstrating a commitment to preserving individual privacy rights on a worldwide scale.

9.1.7 Public Involvement and Collaboration

Ethical AI in privacy requires not only organizational accountability but also active interaction with the public and collaboration with stakeholders. Soliciting public feedback, embracing multiple perspectives, and including persons in decision-making processes all contribute to the ethical deployment of AI systems. Open communication builds trust and ensures that privacy considerations reflect the values and expectations of the communities affected by AI technologies.

In conclusion, as enterprises harness the revolutionary power of artificial intelligence, ethical issues of AI in privacy are critical. Balancing innovation and privacy, assuring informed consent, addressing bias, preserving anonymity, limiting intrusiveness, fostering responsibility, adhering to global norms, and involving the public are all important components of ethical AI practices. By carefully managing these considerations, enterprises can not only embrace the potential of AI but also contribute to a future in which innovation and privacy coexist together.

9.2 The Role of AI Ethics in Data Handling and Privacy Protection

The ethical considerations around data processing and privacy protection have become critical in the dynamic world of artificial intelligence (AI). As businesses increasingly use AI technologies to process and generate insights from massive datasets, the ethical consequences of these practices become clear. This chapter examines the ideas and methods that guide ethical use of AI in the world of personal information, as well as the critical role of AI ethics in data handling and privacy protection.

9.2.1 Ethical Principles as a Base

The application of a set of fundamental principles lies at the heart of AI ethics in data handling. These values, which are frequently based on openness, justice, accountability, and respect for individual rights, form the ethical foundation that governs AI system development and deployment. Organizations dedicated to ethical data handling prioritize these principles, ensuring that every stage of the data lifecycle—from collection to processing to storage—is ethically compliant.

9.2.2 Data Practices Transparency

Transparency is essential in AI ethics, especially when dealing with sensitive data. Organizations must be open and honest about their data practices, disclosing precisely how data is gathered, processed, and used. Transparent methods foster trust among users and stakeholders, allowing them to make informed decisions about disclosing personal information. This transparency includes not just the technical features of AI algorithms, but also the goals for which the data is used, encouraging an open and accountable culture.

9.2.3 Data Processing Fairness

Fairness in data processing is an ethical need for AI applications. Inadvertent or systemic bias in algorithms can result in discriminatory outputs, exacerbating existing disparities. Ethical data management necessitates firms actively identifying and mitigating biases in AI models.

This entails using diverse and representative datasets, assessing algorithms on a regular basis for fairness, and establishing ways to remedy any discrepancies that develop during data processing.

9.2.4 User Control and Consent

Individual autonomy is a critical component of AI ethics in data handling. Before collecting and processing data from users, organizations must seek their informed consent. The consent procedure should be straightforward and explicit, with users having granular control over what data is gathered and how it is utilized. Ethical data management entails giving consumers control over their privacy preferences, such as the ability to opt in or out of data processing activities based on their preferences.

9.2.5 Designing for Security and Privacy

The concept of privacy by design emphasizes the incorporation of privacy safeguards into the AI system development process from the start. Ethical data management necessitates enterprises prioritizing security and implementing effective privacy controls throughout the data lifecycle. This includes secure data storage, encryption, and adherence to cybersecurity best practices. Organizations can proactively limit potential privacy issues by incorporating privacy considerations into the design and architecture of AI systems.

9.2.6 Practices for Responsible Data Sharing

Responsible data sharing is an important part of AI ethics in today's linked digital landscape. Organizations must establish a balance between data sharing for collaborative objectives and individual privacy protection. When sharing insights or cooperating on research, ethical data handling entails using data anonymization and aggregation procedures.

Data sharing techniques that are responsible ensure that the benefits of AI collaboration are realized without jeopardizing the privacy of persons whose data is involved.

9.2.7 Auditing and Continuous Monitoring

Continuous monitoring and auditing of AI systems is required for ethical data handling. Organizations must evaluate the effectiveness of their algorithms on a regular basis in order to discover and correct any concerns relating to privacy or ethical considerations. This entails performing regular privacy impact assessments, algorithmic audits, and remaining up to date on evolving best practices in AI ethics. Continuous monitoring guarantees that businesses can respond to changing ethical norms and new difficulties as they occur.

9.2.8 AI Ethics Training and Education

Fostering a culture of awareness and education inside organizations is a vital part of safe data handling. Employees participating in AI development and data management must be educated on ethical issues, privacy rules, and the potential impact of their work on individuals. Ethical data procedures are not static; they necessitate continual education to keep teams up to date on the most recent ethical standards and best practices in the ever-changing field of AI.

AI and the Future of Privacy: A Delicate Dance

The complex interplay between AI and data privacy necessitates ethical vigilance. This chapter looked at how core principles like transparency, fairness, and user control lay the groundwork for responsible AI development. By prioritizing these principles, organizations can foster a future in which AI innovation thrives alongside strong data privacy safeguards. Continuous collaboration among stakeholders (industry, policymakers, and the general public) is essential for navigating the changing landscape of AI ethics. Finally, a commitment to ethical data practices will ensure that AI is a positive force, empowering individuals and driving innovation in a privacy-conscious world.

CHAPTER 10

AI and Data Security

10.1 AI-Driven Data Security Strategies

In the fast-changing cybersecurity landscape, enterprises are increasingly turning to artificial intelligence (AI) to strengthen their data security strategy. This section delves into the underlying concepts of AI-driven data security policies, studying how AI technologies play a critical role in strengthening organizations' resilience against an ever-expanding array of cyber threats.

10.1.1 Traditional Data Security Measures Overview

For many years, traditional data security methods depended on rule-based systems, firewalls, and signature-based intrusion detection systems. While effective to some extent, these tactics fail to keep up with the ever-changing nature of modern cyber threats. Organizations are recognizing the need for innovative, adaptive security solutions as the digital world grows more interconnected and intelligent.

© Ranadeep Reddy Palle, Krishna Chaitanya Rao Kathala 2024
R. R. Palle and K. C. R. Kathala, *Privacy in the Age of Innovation*,
https://doi.org/10.1007/979-8-8688-0461-8_10

10.1.2 The Impact of AI on Data Security

The ability of AI to analyze massive volumes of data, discover trends, and respond in real time to emerging risks has had a transformative influence on data security. A subset of AI, machine learning algorithms, excels at detecting abnormalities and potential security concerns in large datasets. This dynamic capacity enables enterprises to move beyond static security measures and adopt a proactive and responsive approach to protecting critical data.

10.1.3 Using AI to Detect Threats Proactively

The ability to detect preemptive threats is one of the primary features of AI-driven data security strategies. Traditional approaches sometimes rely on predetermined rules, leaving them open to innovative and developing attack vectors. AI, on the other hand, learns from past data, always improving its grasp of normal and deviant behavior. This helps firms to detect anomalies that indicate cyberattacks, offering an early warning system that can mitigate any damage before it occurs.

10.1.4 Behavioral Analysis and Detection of Anomalies

A key component of AI-driven data security is behavioral analysis, which involves monitoring user and system behavior to spot departures from established norms. AI systems can examine user actions, network traffic, and system logs for unexpected patterns that may indicate a security breach. Organizations can increase their entire security posture by using anomaly detection algorithms to quickly identify and respond to illegal access, data exfiltration, or other malicious activity.

10.1.5 Response Orchestration and Automation

Not only does AI excel at detection, but it also provides enterprises with automatic response capabilities. When confronted with a cyber threat, AI-driven systems can perform specified actions such as isolating affected systems, preventing suspicious activity, or even launching incident response processes. Automation shortens response times, allowing firms to detect possible attacks in near real time and decrease the impact of security breaches.

10.1.6 Adaptation and Continuous Learning

In the ever-changing landscape of cyber threats, AI's adaptability is a vital tool. AI systems continuously learn from new data, updating their models to evolving cyber adversary strategies. This ability to learn on the fly allows organizations to keep ahead of emerging threats, giving a level of agility that traditional security solutions fail to match.

Finally, AI-driven data security techniques reflect a paradigm shift in how businesses approach cybersecurity. Organizations may develop resilient defenses against an increasingly sophisticated threat landscape by leveraging the power of AI for proactive threat detection, behavioral analysis, and automated response orchestration. As the digital ecosystem evolves, incorporating AI in data security is no longer a strategic choice, but a requirement for enterprises devoted to protecting their sensitive data.

10.2 Adversarial Machine Learning in Security

Adversarial machine learning has emerged as a double-edged blade in the complex dance between cybersecurity and technological innovation. This section delves into the complex world of adversarial machine learning in the context of data security, revealing both the potential vulnerabilities it creates and the countermeasures required to protect AI systems.

10.2.1 How Attackers Target AI System Vulnerabilities

Adversarial attacks take use of AI models' inherent vulnerabilities throughout the learning process. In order to insert tiny perturbations into input data, attackers use comprehensive knowledge of the model's architecture and training data. These perturbations are deliberately designed to deceive the AI system while remaining undetectable to human observers. Adversaries intend to exploit the model's flaws and create misclassifications, which could lead to security breaches.

10.2.2 Defense and Countermeasure Strategies

To mitigate the hazards posed by adversarial machine learning, a multifaceted approach is required. A strong security framework must include strong counters and defense techniques.

- Adversarial Practice: One important method is to include hostile examples in the training data. When AI models are exposed to contrived adversarial inputs during training, they learn to recognize and reject such manipulations, increasing their resistance in real-world circumstances.

- Ensemble Methods: By combining many models into an ensemble, a collective defense against adversarial attacks can be provided. Adversaries find it difficult to design attacks that are effective across multiple models, making ensemble approaches a powerful defense mechanism.

- Regularization Methods: Regularization techniques used throughout the training process serve to prevent overfitting to certain patterns, making the model more resistant to hostile manipulation. Dropout and weight regularization techniques improve the generalization capabilities of AI models.

- Anomaly Detection: By incorporating anomaly detection algorithms, unexpected patterns in input data can be identified. Anomaly detection systems can trigger alerts or preventive measures when an AI system meets adversarial inputs that differ considerably from regular patterns.

- Adversarial Detection Systems: Adding an additional layer of defense by implementing specialized systems designed to detect adversarial attacks in real time. Anomaly detection, statistical analysis, and heuristics are used by these systems to identify and mitigate potential hostile threats.

10.2.3 Adversarial Machine Learning Ethical Considerations

Aside from technological concerns, the ethical aspect of adversarial machine learning cannot be disregarded. As organizations establish defense systems, ethical considerations must govern countermeasure deployment. To minimize unforeseen repercussions and assure responsible AI operations, it is critical to strike a balance between solid security and ethical usage of adversarial approaches.

10.2.4 In the Future: Evolving Threats and Adaptive Defenses

Adversarial machine learning is a developing field in which attackers and defenders play a never-ending cat-and-mouse game. Adversarial approaches become more complicated as AI models become more smart. The creation of adaptive defenses that can grow alongside evolving adversarial threats is the future of data security. Continuous research, collaboration, and a proactive approach are required to keep one step ahead in this ever-changing landscape.

Finally, adversarial machine learning brings a complex layer of problems to data security. Organizations, on the other hand, can strengthen their AI systems with strategic countermeasures, ensuring they remain strong and resilient in the face of evolving adversarial threats. The route to safeguarding AI systems is more than just a technical exercise; it also requires a commitment to ethical standards that promote responsible and secure AI deployment.

10.3 Emerging Trends in AI and Data Security

The landscape of data analytics and data security is always evolving, molded by technology developments and the ever-changing strategies of cyberattackers. This section digs into the developing trends that are transforming the future of data security, including creative tactics that use artificial intelligence to keep ahead of evolving threats.

10.3.1 AI and Blockchain Collaboration for Secure Transactions

As blockchain technology gains traction, the combination of AI and blockchain emerges as a powerful force in strengthening data security. Blockchain's decentralized and immutable nature compliments AI's capabilities in establishing transparent, tamper-resistant ledgers. This collaboration not only maintains transactional integrity but also adds additional dimensions to safe data sharing and authentication, particularly in industries such as finance and healthcare.

10.3.2 Quantum Computing and Its Implications for Data Security

While quantum computing promises unrivaled computational capability, it also poses a potential challenge to current encryption systems. As quantum computers progress, they may be able to break traditional encryption systems. As a result, academics are investigating quantum-resistant cryptographic algorithms, and AI is playing an important role in building and modifying security protocols that may withstand the potential vulnerabilities posed by quantum computing.

10.3.3 AI Integration in Zero-Trust Security Models

The zero-trust security model, formerly considered a radical concept, is gaining momentum as a conventional approach to data protection. Zero-trust security models use the "never trust, always verify" approach. They remove implicit trust from a network and continually authenticate all users and devices attempting to access resources. Building zero-trust systems entails enforcing deny-all by default policies, granting least privilege access based on context, and continuously monitoring user behavior for anomalies. AI can improve zero-trust security by analyzing user behavior patterns and detecting suspicious activities that traditional rule-based systems might miss. This paradigm challenges the traditional notion of trusting entities within a network and instead calls for constant verification of every user, device, or program entering the network. AI can help create and improve zero-trust infrastructures by delivering real-time threat intelligence, behavioral analytics, and automated response mechanisms.

10.3.4 Autonomous Security Systems

The introduction of autonomous security systems powered by AI is a game changer in data security. These systems can dynamically react to evolving threats, evaluate massive databases in real time, and make fast judgments to mitigate any security breaches. Autonomous security systems shorten response time to cyber threats, resulting in a more agile defense mechanism capable of keeping up with the speed and complexity of modern cyberattacks.

10.3.5 Explainable AI in Security Decision-Making

The requirement for transparency and accountability in AI decision-making is pushing the use of explainable AI in data security. Understanding how AI systems arrive at conclusions is critical as these technologies become more integrated into security operations. Explainable AI delivers insights into the decision-making processes of AI algorithms, allowing security professionals to interpret and trust the outcomes while also aiding regulatory compliance.

10.3.6 AI-Driven Threat Hunting and Intelligence

Traditional threat detection approaches sometimes rely on preset patterns, leaving firms open to evolving attacks. AI-driven threat hunting use machine learning to proactively seek out potential risks by evaluating massive datasets for subtle indicators of compromise. This tendency enables firms to keep ahead of sophisticated attackers and detect innovative attack channels before they cause significant damage.

10.3.7 Human Augmentation in Security Operations

The merging of AI with human expertise, known as human augmentation, is altering security operations. AI supports security analysts by automating basic activities, analyzing vast datasets, and discovering patterns, allowing human experts to focus on strategic decision-making and addressing difficult security concerns. This collaborative method improves overall security effectiveness.

Finally, the future of AI and data security is distinguished by a convergence of novel trends that employ artificial intelligence to handle rising dangers. The convergence of AI and blockchain, the influence of quantum computing, the emergence of zero-trust models, autonomous security systems, explainable AI, AI-driven threat hunting, and human augmentation all define the cutting edge of data security. By embracing these trends, companies can traverse the growing threat landscape with resilience and foresight, assuring effective protection for their sensitive data.

CHAPTER 11

Balance Between Security and Privacy

11.1 Exploring the Trade-Off Between Security and Individual Privacy

The delicate balance between security and individual privacy emerges as a prominent theme in the complicated tapestry of digital living. This section looks into the complicated dynamics of the trade-off between security imperatives and individual privacy preservation, revealing the complexities and concerns that define this delicate balance.

11.1.1 The Security and Privacy Paradox

The pursuit of effective security frequently entails the acquisition and analysis of huge volumes of personal information, which is at the heart of the security-privacy conundrum. In the attempt to defend digital fortresses against a wide range of cyber dangers, corporations may find themselves wrestling with the ethical issues of data collection, accidentally invading individuals' privacy.

© Ranadeep Reddy Palle, Krishna Chaitanya Rao Kathala 2024
R. R. Palle and K. C. R. Kathala, *Privacy in the Age of Innovation*,
https://doi.org/10.1007/979-8-8688-0461-8_11

11.1.2 Navigating the Regulatory Environment

Navigating the regulatory landscape has become a vital component of the security-privacy equation in an era of increased awareness about data protection. Regulations such as the General Data Protection Regulation (GDPR) and the California Consumer Privacy Act (CCPA) establish strict guidelines for the collecting, processing, and storage of personal information. Organizations must find a difficult balance between meeting legislative requirements to safeguard privacy and strengthening their security measures.

11.1.3 Data Reduction and Purpose Restrictions

The concept of data reduction is a crucial principle in addressing the security-privacy trade-off. Organizations should acquire just the personal data required for a specified purpose, minimizing needless invasions of individuals' privacy. Purpose limitation supports this by highlighting that data should only be utilized for the purpose for which it was gathered, limiting the potential for security overreach.

11.1.4 User Empowerment and Consent

In the delicate balance between security and privacy, obtaining informed consent from individuals emerges as a cornerstone. Transparency regarding data collecting procedures and the exact goals for which data will be used enables users to make educated decisions about providing their personal information. Users should have granular control over consent processes, allowing them to modify their choices and strike the optimal balance between security and privacy.

11.1.5 Security Solutions with Privacy by Design

Integrating privacy by design principles into security solutions is a proactive strategy for reducing the security-privacy trade-off. Organizations can ensure data protection is an essential part of their entire strategy by incorporating privacy considerations into the architecture and development of security solutions. This method promotes a peaceful cohabitation in which security is robust and privacy is protected by design.

11.1.6 Encryption as a Privacy Protection Measure

Encryption appears as a potent ally in the arsenal of privacy protection technologies. Strong encryption technologies not only secure data during transmission and storage, but they also help to protect individual privacy. Organizations can strengthen security without jeopardizing personal information confidentiality by using end-to-end encryption and secure communication protocols.

11.1.7 Finding the Right Balance: A Lifelong Challenge

The pursuit of a balance between security and privacy is an ongoing task that necessitates constant reevaluation. Organizations must modify their methods to maintain both robust security and unwavering privacy protection as technology advance and new threats emerge. Achieving this delicate balance is an ongoing process of refinement and optimization, not a one-time endeavor.

11.1.8 Public Attitudes and Trust

Public trust is heavily influenced by how organizations handle the security-privacy trade-off. Transparency, ethical data practices, and a dedication to user rights all contribute to the establishment and maintenance of confidence. Organizations that emphasize privacy concerns with security measures are more likely to develop a positive user perception.

To summarize, the trade-off between security and individual privacy is a difficult and ever-changing challenge. To find a delicate balance, organizations must manage legislative obligations, implement privacy-centric design principles, seek informed permission, and employ encryption technology. By taking these factors into account, enterprises may create an atmosphere in which security and privacy coexist peacefully, earning the trust of individuals in the digital realm.

11.2 Navigating the Regulatory Landscape

In the ever-changing digital era, the regulatory landscape is critical in shaping the delicate balance between security imperatives and individual privacy rights. This section delves into the complicated web of legislation governing data protection, investigating how firms negotiate this treacherous terrain to achieve compliance while bolstering their security procedures.

11.2.1 Striking a Balance: Compliance and Security Measures

Navigating the regulatory landscape necessitates enterprises striking a fine balance between completing compliance standards and bolstering their security measures. While legislation requires enterprises to adhere to particular data protection principles, organizations must also fight with the ever-changing nature of cyber threats, needing sophisticated security protocols to preserve sensitive information.

11.2.2 Data Minimization and Purpose Limitation

Key principles incorporated in data protection rules, such as GDPR, argue for data minimization and purpose limiting. These principles emphasize the necessity of collecting only the least amount of personal data required for a certain purpose. By adhering to these principles, corporations not only comply with legislation but also contribute to the preservation of individual privacy, which aligns with the broader goals of data protection laws.

11.2.3 Consent Mechanisms and User Rights

The implementation of appropriate consent processes and the respect of user rights are critical for navigating the regulatory landscape. Regulations highlight the significance of gaining informed consent from individuals before collecting and using their personal data. Transparent communication about data practices, user-friendly consent interfaces, and giving consumers control over their data all contribute to compliance and establish a trusting connection between organizations and users.

11.2.4 Accountability and Privacy Impact Assessments

Privacy impact assessments (PIAs) are a method that regulations propose for assessing the potential impact of data processing activities on person privacy. Organizations can detect and reduce risks by conducting PIAs, ensuring that their security measures correspond with privacy objectives.

Furthermore, regulations emphasize accountability as a basic tenet, requiring firms to demonstrate their commitment to compliance through documentation, transparency, and responsible data practices.

11.2.5 Global Compliance and International Data Transfers

Organizations must handle the issues of international data transfers in an interconnected world where data travels seamlessly across borders. Adhering to frameworks like the EU-US Privacy Shield or creating enforceable corporate norms makes global data compliance easier. Organizations that operate in many jurisdictions must manage the complexities of regional requirements while maintaining a united commitment to data protection.

11.2.6 Changing Regulations and Constant Adaptation

The regulatory landscape is dynamic, with new legislation being introduced to meet developing concerns. Organizations must be agile in adapting to these changes, staying up to date on advances in data protection rules and updating their security procedures accordingly.

11.2.7 Ethical Considerations Beyond Compliance

While regulatory compliance is critical, firms committed to successfully navigating the regulatory landscape also address broader ethical implications. Maintaining ethical data practices goes beyond legal requirements, cultivating a culture of trust and accountability that goes beyond regulatory compliance.

To summarize, managing the regulatory landscape is a multifaceted task that necessitates enterprises harmonizing security measures with the specific needs of data protection rules. By embracing principles such as data minimization, gaining informed consent, completing privacy effect assessments, and demonstrating accountability, enterprises can not only comply with rules but also contribute to the larger objective of safeguarding individual privacy in the digital age.

CHAPTER 12

Best Practices and Recommendations

12.1. Guidelines for Organizations to Implement AI for Enhanced Security and Privacy

As enterprises use artificial intelligence (AI) to improve security and privacy, a thorough set of rules becomes essential. This section outlines fundamental ideas and best practices that businesses may use when deploying AI to provide increased security and privacy in the digital realm.

12.1.1 Designing Holistic Security and Privacy

Adopting a holistic strategy is the foundation of a successful AI implementation for increased security and privacy. Integrating security and privacy issues into the design and development phases, referred to as security and privacy by design, guarantees that these principles are at the heart of AI systems. This proactive technique reduces the chance of neglecting crucial features later in the development process and aligns the process with ethical and regulatory norms.

12.1.2 Carrying Out Privacy Impact Assessments (PIAs)

For firms looking to use AI responsibly, privacy impact evaluations are invaluable tools. Conduct complete PIAs before deploying AI systems to detect and mitigate potential privacy problems. These studies enable enterprises to understand the implications of their AI initiatives on individual privacy, allowing them to make educated decisions and mitigate risk.

12.1.3 Openness and Explainability

Transparent AI solutions help to build user confidence and promote ethical data practices. Transparency in describing how AI is used for security and privacy should be prioritized by organizations. Furthermore, guaranteeing explainability in AI models—that is, making the decision-making process intelligible to stakeholders—builds trust and helps users to understand how their data is being utilized, aligning with openness and accountability principles.

12.1.4 Data Reduction and Purpose Restrictions

In the AI landscape, adhering to the principles of data minimization and purpose limitation is crucial. Collect only the personal data needed for the intended purpose, avoiding the danger of unnecessary data exposure. Define the objective of data collection and processing clearly, while adhering to regulatory standards and protecting individual privacy.

12.1.5 Robust Data Security Measures

To protect against cyber threats, AI implementations should be backed by robust data security measures. To protect sensitive information, use encryption, access controls, and secure data storage techniques. Updating security rules on a regular basis to meet evolving threats ensures that AI systems are resilient in the face of potential breaches.

12.1.6 User Empowerment and Informed Consent

Prioritize getting users' informed consent before collecting and using their data. Create consent systems that are visible, simple to comprehend, and provide users granular control over their data. Giving people the ability to make informed decisions about their privacy preferences increases control and coincides with ethical data practices.

12.1.7 Auditing and Continuous Monitoring

Create a solid mechanism for monitoring and auditing AI systems on a constant basis. Evaluate the effectiveness of AI algorithms on a regular basis to identify and correct any security or privacy issues. This continuous method ensures that businesses remain alert to emerging difficulties and can react to changing ethical standards and best practices.

12.1.8 Employee Education and Awareness

Ensure that staff participating in AI development and data handling receive comprehensive training on best practices for security and privacy. Develop an organizational culture of awareness and education that emphasizes the ethical considerations connected with AI implementation. Teams that are well-informed are better positioned to make decisions that favor security and privacy.

12.1.9 Stakeholder Collaboration

Engage in open communication and collaboration with stakeholders such as users, regulators, and peers in the sector. Solicit feedback from appropriate stakeholders to continuously develop AI systems. Organizations can develop trust, resolve concerns, and demonstrate a commitment to responsible and ethical AI techniques by encouraging open discourse.

12.1.10 Adaptability and Flexibility

Recognize that the AI landscape is fluid and that technology advances quickly. Implement adaptable frameworks that can accommodate regulatory changes, emerging dangers, and technology developments. A proactive approach to adaptation guarantees that enterprises can effectively traverse the developing landscape of AI, security, and privacy.

Lastly, successful AI adoption for better security and privacy necessitates a complete set of rules that include ethical concerns, regulatory compliance, and robust technical protections. By implementing these principles, businesses can not only strengthen their security posture but also provide the groundwork for trust and accountability in their usage of AI technologies.

Case Study: Implementing AI for Enhanced Security and Privacy at Acme Inc.

Acme Inc., a leading financial services company, recognized AI's potential to enhance customer and organizational security and privacy. They were also aware of the inherent risks associated with artificial intelligence, such as data breaches and biased algorithms. To address these concerns, Acme implemented a multipronged approach of using AI responsibly:

1. Techniques for Anonymizing Data and Protecting Privacy: Acme anonymized customer data before using it in AI models. This involved using techniques such as tokenization and differential privacy to keep customer information private while still allowing AI to extract valuable insights for fraud detection and risk management.
   For tasks that required some level of identification, Acme used federated learning. AI models were trained on decentralized datasets owned by individual branches, which reduced data sharing and increased privacy.

2. Securing the AI Development Lifecycle: Acme implemented a secure AI development lifecycle (SAIDLC) framework similar to Google's Secure AI Framework (SAIF) `https://blog.google/technology/safety-security/introducing-googles-secure-ai-framework/`. This framework emphasized secure coding practices, vulnerability assessments, and ongoing monitoring of AI models in order to detect and mitigate potential security risks. Regular penetration testing of AI systems ensured their resilience to cyberattacks.

3. Explanatory AI (XAI): Acme used XAI techniques to understand how AI models make decisions. This transparency enabled them to detect and correct potential biases in the algorithms, promoting

fairness and preventing discriminatory outcomes. Acme built trust with customers and regulators by explaining AI-driven loan approval and fraud detection decisions.

4. Government and Oversight: Acme formed a dedicated AI governance committee comprised of security, privacy, and ethics experts. This committee oversaw AI development and deployment, ensuring adherence to regulations and ethical principles.

    Regular employee training sessions on responsible AI practices enabled employees to identify and report potential risks.

Results: Acme's security and privacy improved significantly after implementing these best practices:

- AI's ability to identify complex patterns increased fraud detection rates by 20%.

- Data breaches were minimized by implementing strong security measures throughout the AI lifecycle.

- Acme's commitment to data privacy and explainable AI helped to build customer trust.

Conclusion: Acme's case study shows how organizations can use AI to improve security and privacy. Organizations can reap the benefits of AI while minimizing associated risks by implementing a responsible AI approach that prioritizes data anonymization, secure development, explainability, and strong governance.

12.2. Ensuring Responsible and Ethical AI Practices

Responsible and ethical standards are critical in the age of artificial intelligence (AI) for establishing trust, reducing risks, and guaranteeing the long-term success of AI initiatives. This section highlights critical suggestions for businesses to follow when using AI, highlighting the need of accountability and integrity in the development and deployment of AI systems.

12.2.1 Ethical Issues in AI Development

The cornerstone of responsible AI activities begins during the development process with ethical considerations. Organizations should prioritize transparency in AI model development by publicly disclosing algorithm design, data selection, and training decisions. Organizations may design ethical AI systems by minimizing biases, maintaining diversity in training datasets, and addressing potential ethical issues.

12.2.2 Fairness and Bias Avoidance

It is critical to address biases in AI systems in order to maintain fairness and prevent discriminatory consequences. To discover and correct biases in training data, use fairness-aware machine learning approaches. Audit and review AI models on a regular basis to ensure that diverse demographic groups are treated fairly. This dedication to fairness is consistent with ethical principles and encourages inclusion in AI applications.

12.2.3 Accountability and Explicitness

Accountability and explainability are critical components of responsible AI. Organizations should define roles and responsibilities for engineers, data scientists, and decision-makers when it comes to AI systems. Additionally, stress explainability in AI models to ensure that algorithmic decisions are interpretable and understandable. This transparency fosters confidence and allows people to understand the reasons behind AI-generated results.

12.2.4 User-Centric Design and by Default Privacy

Adopting a user-centered design approach puts people at the center of AI solutions. Create systems that prioritize user privacy, giving people control over their data and preferences. Implement privacy by default, which means that privacy safeguards are built into the design and functionality of AI systems. This user-centric mindset promotes agency and respect for individuals' privacy rights.

12.2.5 AI Teams' Ongoing Ethical Training

Make sure that AI development teams receive continual ethical training to keep up with changing ethical standards and best practices. This continual education builds an organizational culture of accountability and understanding, helping teams to make ethical judgments at every level of the AI lifecycle. Ethical training should cover not only technical aspects of AI but also the broader societal ramifications of the technology.

12.2.6 Ethical Review Boards and External Audits

Consider hiring external auditors or forming ethical review boards to analyze AI systems objectively. External audits help to increase openness and provide an unbiased assessment of ethical aspects. Ethical review boards comprised of varied experts can provide insights into potential ethical issues and provide mitigations, enabling a comprehensive and unbiased examination.

12.2.7 Impact Assessment and Responsible Deployment

Before deploying AI systems in real-world situations, prioritize responsible deployment methods by completing impact assessments. To predict and prevent unexpected repercussions, assess the potential societal, ethical, and privacy ramifications of AI applications. Responsible AI deployment entails actively monitoring the impact of AI systems on individuals and communities and making required improvements.

To end, establishing responsible and ethical AI processes is a strategic need, not just a moral one. Organizations may create trust, manage risks, and contribute to a good societal impact by including ethical considerations into all stages of AI research and implementation. These approaches not only adhere to ethical guidelines but also position enterprises as responsible stewards of AI technology in a world that is becoming increasingly linked and data-driven.

CHAPTER 13

Future Trends and Challenges

13.1 Emerging Trends in AI, Information Security, and Data Privacy

A dynamic landscape of rapid technological developments and increasing challenges shapes the future of AI, information security, and data privacy. This section delves into the emerging developments that will reshape the convergence of AI, information security, and data privacy in the coming years.

13.1.1 AI and Cybersecurity Convergence

The increasing convergence of AI and cybersecurity is a key trend on the future. AI technologies, particularly machine learning, are becoming increasingly important in bolstering cybersecurity defenses. Artificial intelligence-powered threat detection, behavioral analysis, and automated response systems are evolving to predict and defeat sophisticated cyber threats, ushering in a new era of proactive and adaptive cybersecurity.

R. R. Palle and K. C. R. Kathala, *Privacy in the Age of Innovation*,
https://doi.org/10.1007/979-8-8688-0461-8_13

13.1.2 AI Technologies That Protect Your Privacy

As worries about data privacy grow, so does the development of privacy-preserving AI solutions. Techniques like homomorphic encryption, safe multiparty computation, and federated learning enable enterprises to utilize the power of AI while protecting sensitive data. These developments indicate a move toward balancing the dual goals of innovation and data protection.

13.1.3 Responsible AI and Ethical AI

The ethical dimension of AI is becoming increasingly important in the development and deployment of AI systems. Responsible innovation necessitates a multifaceted strategy that takes into account not only technical concerns but also the larger societal implications of AI applications. Ethical AI frameworks, guidelines, and certifications are anticipated to gain traction, directing enterprises toward ethical and accountable AI practices.

13.1.4 Quantum Security Cryptography

The cryptography landscape is set to change with the advent of quantum computing. Quantum-safe cryptography is a new movement that aims to create encryption solutions that are resistant to quantum attacks. Organizations will need to adopt quantum-safe encryption methods as quantum computers evolve to assure the continuous security of their data in the post-quantum era.

13.1.5 AI That Can Be Explained for Trust and Accountability

Explainability in AI is gaining popularity as a technique of increasing trust and accountability. As AI systems make decisions that affect people and society, the ability to explain how these decisions are made becomes increasingly important. Explainable AI approaches, such as interpretable machine learning models, are expected to play an important role in providing transparency and building trust in AI applications.

13.1.6 Blockchain and AI Integration

The confluence of blockchain and AI has the potential to transform data integrity and security. The decentralized and tamper-resistant structure of blockchain supplements AI's capabilities by providing a transparent and immutable ledger for documenting AI-generated insights. This connection has the potential to improve data traceability, reduce the risk of manipulation, and contribute to a more secure and accountable data ecosystem.

13.1.7 Global Standards and Regulatory Evolution

The regulatory environment surrounding AI, information security, and data privacy is projected to change dramatically. To address the ethical, privacy, and security concerns of AI technologies, governments and international agencies are likely to propose new legislation and standards. The development of global norms and standards will be critical in promoting responsible AI techniques on a larger scale.

Finally, the future of AI, information security, and data privacy is defined by a convergence of scientific advancements, ethical issues, and growing legislative frameworks. Organizations that remain on top of these developing trends will be better positioned to negotiate the intricacies of this landscape, adopting responsible and innovative ways that balance the imperatives of security, privacy, and technical innovation.

13.2 Predictions and Potential Challenges

As AI, information security, and data privacy continue to improve, the future holds both promise and difficulty. This section dives into the forecasts and probable difficulties that will influence the landscape in the future years, providing insights into the trajectory of various interconnected fields.

13.2.1 Prediction: Explainable AI Advances

Continued developments in Explainable AI are one prediction on the horizon (XAI). The demand for transparency in decision-making processes grows as AI systems become more complex and prevalent. Explainable AI techniques, which allow people to comprehend how AI arrives at specific conclusions, are predicted to advance. The increased emphasis on responsibility, regulatory compliance, and the user's right to understand and question AI-driven judgments is driving this trend.

13.2.2 Forecast: Greater Focus on AI Ethics and Governance

In the coming years, ethical concerns about AI are likely to take center stage. Organizations and policymakers will place a greater emphasis on developing strong AI ethics frameworks and governance systems.

Addressing biases in AI algorithms, guaranteeing fairness, and creating clear criteria for the responsible and ethical use of AI technologies are all part of this. The goal is to strike a balance between innovation and individual rights protection.

13.2.3 Forecast: The Rise of AI-Driven Cybersecurity Threats

As AI becomes more integrated into cybersecurity protection, the emergence of AI-driven cybersecurity threats is predicted. Adversarial machine learning, in which attackers utilize AI to exploit security system flaws, is predicted to become more advanced. To remain ahead of AI-driven dangers, organizations will need to constantly improve their cybersecurity safeguards, ushering in an era of AI-driven cyber-physical risks.

The ever-changing threat landscape necessitates proactive security strategies. The concepts discussed in this book, such as zero-trust architectures and secure design techniques, enable organizations to stay ahead of emerging threats. Zero-trust minimizes network trust by continuously verifying access requests. Secure design principles prioritize security throughout the development process, increasing resilience to novel attacks. Organizations that adopt these approaches can strengthen their defenses and adapt to the ever-changing security landscape.

13.2.4 Forecast: AI Integration in Personalized Privacy Solutions

One noteworthy forecast is the incorporation of AI in individualized privacy solutions. As people become more aware of their digital footprint, there is an increasing desire for customizable privacy options. AI-driven systems that adapt to customers' preferences while providing granular control over data sharing and privacy settings are predicted to grow in popularity. This approach is consistent with the larger trend of enabling individuals to regulate their privacy in an increasingly connected society.

13.2.5 Difficulty: Regulatory Fragmentation and Compliance Complicatedness

The increasing complexity of regulatory landscapes is one potential difficulty that enterprises may confront. As more nations and regions implement AI-specific legislation and data protection laws, firms operating on a global scale may face difficulties navigating varied and sometimes conflicting compliance requirements. Harmonizing compliance efforts and adjusting to changing regulatory regimes will be key future challenges. While navigating the complex landscape of data privacy regulations can be difficult due to regional variations, there is a growing push for international harmonization. Organizations like the Global Privacy Assembly are working to unify core data privacy principles and establish interoperable frameworks. This potential convergence could simplify compliance for international businesses and, in the long run, benefit global user privacy.

13.2.6 Task: Ethical Application of AI in Decision-Making

The ethical application of AI in decision-making processes is a possible challenge. As AI systems assume greater decision-making responsibilities in numerous industries, ensuring that these judgments adhere to ethical norms becomes increasingly important. To avoid unforeseen effects and ethical failures, businesses must strike the correct balance between automated decision-making and human oversight.

13.2.7 Difficulty: Addressing Bias and Fairness

Managing bias and maintaining fairness in AI systems is an ongoing challenge. Biases in training data and algorithmic decision-making can result in biased outcomes, despite continued attempts. Overcoming this difficulty will necessitate not only technical answers but also a commitment to diversity and inclusivity, as well as continual monitoring of AI systems for potential biases.

To finish, the future of artificial intelligence, information security, and data privacy contains both exciting opportunities and complicated challenges. Organizations that anticipate these trends and handle possible difficulties ahead of time will be better positioned to traverse the developing landscape, enabling responsible, ethical, and safe AI technology deployment.

Conclusion

Summarizing Key Takeaways

A plethora of insights have been explored as part of the journey through "Enhancing Information Security, Data Privacy, and Data Security with AI: Balancing Innovation and Privacy," shedding light on the intricate interplay between artificial intelligence (AI), information security, and data privacy. As we come to the end of this illuminating journey, it's critical to extract the essential conclusions that capture the spirit of the discussion.

Integrating AI and Security

The incorporation of artificial intelligence into information security emerges as a transformational force. The ability of AI to identify, mitigate, and provide cybersecurity solutions highlights its critical role in bolstering digital defenses. Organizations may proactively tackle changing cyber threats by integrating AI with security protocols, proving the symbiotic relationship between innovation and digital asset protection.

AI Techniques for Protecting Privacy

Techniques like homomorphic encryption, differential privacy, and safe multiparty computation underscore the importance of preserving data privacy in the midst of the AI revolution. These privacy-preserving solutions enable enterprises to leverage the power of AI without

R. R. Palle and K. C. R. Kathala, *Privacy in the Age of Innovation*,
https://doi.org/10.1007/979-8-8688-0461-8

jeopardizing individual privacy. Federated learning shines as a beacon, enabling collaborative learning without the use of centralized data and ushering in a new era of privacy-conscious AI applications.

Navigating Regulatory Environments

The regulatory environment, governed by standards such as GDPR and CCPA, acts as a critical framework for enterprises. Compliance with these regulations assures not just legal compliance but also coincides with the ethical requirement of respecting user rights. Navigating this complex legal terrain necessitates a delicate balance, with firms required to follow data protection regulations while employing AI to improve security.

Ethical AI Model Security

The security of AI models extends beyond technological considerations to ethical considerations. Model explainability, fairness, and deployment security are all best practices. Striking this balance ensures that AI applications are not only resistant to threats but also follow ethical norms, fostering trust and accountability in the deployment of AI-driven solutions.

The Cost-Benefit Analysis Security Versus Privacy

The eleventh chapter goes into the complex trade-off between security imperatives and individual privacy. Recognizing this delicate balance is critical, with major issues including data minimization, purpose limitation, and transparent permission processes. Navigating the legal landscape becomes an important component of achieving this balance, underlining the importance of firms embracing responsible and ethical AI techniques.

Prospective Trends and Challenges

Anticipating future trends and difficulties is critical for firms pursuing long-term success. The intersection of AI and cybersecurity, advances in explainable AI, and the incorporation of blockchain all point to a dynamic landscape. Simultaneously, issues such as regulatory complexity, ethical decision-making in AI, and tackling biases present complex obstacles that necessitate proactive answers.

In conclusion, "Enhancing Information Security, Data Privacy, and Data Security with AI: Balancing Innovation and Privacy" serves as a guidebook for traversing the complex terrain of AI, security, and privacy. The primary takeaways emphasize the significance of a comprehensive approach in which technological innovation is aligned with ethical ideals, legal obligations, and the imperative of protecting individual privacy in the digital age. These important lessons serve as guiding principles for establishing the delicate balance between innovation and privacy as firms embark on this transformative journey.

The Future of AI, Information Security, Data Privacy, and Data Security with a Focus on Privacy

As we near the end of "Enhancing Information Security, Data Privacy, and Data Security with AI: Balancing Innovation and Privacy," our attention shifts to the horizon, contemplating the future of AI, information security, and data privacy. This final chapter summarizes the future vision, highlighting the symbiotic relationship between technological innovation and the preservation of individual privacy.

The Evolution of AI: A Change Catalyst

The future of artificial intelligence promises a transformational progression distinguished by improvements that go beyond mere technological competence. AI has the potential to be a good change agent, stimulating innovation in ways that not only improve efficiency and security but also adhere to the ethical ideals of openness, fairness, and responsibility. The convergence of AI and human values will be a defining feature in determining the AI landscape narrative.

AI-Enabled Information Security

As AI advances, information security will be at the forefront of the digital frontier. In the future, AI-driven cybersecurity will become even more sophisticated, predicting and neutralizing threats with remarkable accuracy. The incorporation of AI into security frameworks will not only strengthen digital defenses but will also allow enterprises to remain ahead of the ever-changing threat landscape.

Data Privacy as a Basic Right

In the future narrative, data privacy is viewed as a fundamental right in the digital landscape. Individuals will want greater control and transparency over how their information is used as they become more aware of the fundamental worth of personal data. Privacy-preserving AI solutions will become commonplace, guaranteeing that the benefits of AI innovation are realized without jeopardizing individual privacy.

Ethical Issues in AI Development and Implementation

Ethical considerations will take precedence in the development and deployment of AI. The cornerstone will be responsible innovation, with firms valuing fairness, explainability, and accountability. Ethical review boards, ongoing training, and strong governance structures will be essential components in building a culture in which AI aligns with societal values and positively contributes to human well-being.

The Interaction of Regulatory Frameworks and International Collaboration

Regulatory frameworks governing AI, information security, and data privacy will be refined further, with a focus on a unified approach to global concerns. Collaboration between governments, industry, and regulatory agencies will become increasingly prominent, indicating a shared commitment to developing norms that balance innovation with individual rights protection.

Innovation and Privacy: A Lifelong Journey

The future story depicts a never-ending journey of balancing innovation and privacy. This delicate balance necessitates businesses' ongoing commitment to adapt and evolve. Maintaining the proper balance necessitates not only adherence to regulations and best practices but also an inherent grasp of the ethical implications of technology breakthroughs.

CONCLUSION

To end, the future of artificial intelligence, information security, data privacy, and data security is a story of promise and responsibility. The focus on privacy emerges as a guiding principle as enterprises traverse this dynamic terrain, ensuring that the benefits of technology innovation are utilized responsibly, ethically, and with uncompromising regard for individual privacy rights. The path ahead combines the power of AI with a commitment to crafting a digital future that values both innovation and privacy.

Additional Resources

References and Further Reading

The pursuit of knowledge and mastery in the fields of artificial intelligence, information security, data privacy, and data security is a never-ending path. This chapter provides a portal to a multitude of resources, providing references and further reading materials for readers who want to go deeper into the various fields covered in the book.

Basic Texts and Research Papers

The references section contains foundational texts and research papers that create the foundations for a thorough grasp of artificial intelligence, information security, and data privacy. Esteemed publications by industry thought leaders provide in-depth insights into the theoretical foundations, ethical considerations, and technology breakthroughs that shape the landscape.

Significant Research Journals and Publications

Key research journals and publications are invaluable resources for anyone looking for current research and cutting-edge advancements. These scholarly sources serve as a platform for the distribution of peer-reviewed research, giving readers access to the most recent studies, methodology, and discoveries in artificial intelligence, cybersecurity, and privacy-preserving technology.

Official Guidelines and Standards

Navigating the regulatory landscape and conforming to ethical norms are important components of the ebook's subjects. Authoritarian rules and standards, such as those issued by international organizations and regulatory authorities, are critical resources. These documents lay out best practices, compliance standards, and ethical concerns that businesses should incorporate into their AI and data security policies.

Comprehensive AI and Security Books

Comprehensive guides on AI and security give readers with an immersive understanding of the subject matter in the world of books. These publications cover a wide range of topics, from fundamental AI and machine learning principles to advanced cybersecurity strategies. The choices range from beginner-friendly introductions to expert-level assessments, catering to readers of all skill levels.

Online Courses and Platforms

Online platforms and courses provide a dynamic way to expand one's knowledge for people who prefer interactive learning. Whether it's mastering AI algorithms, deciphering data protection rules, or improving cybersecurity abilities, reliable online platforms and courses offer a flexible and enjoyable learning experience.

Following are some related online resources for further learning:

- Machine Learning for Cybersecurity by Andrew Ng on Coursera: Apply machine learning techniques to detect threats and protect your systems. (`www.coursera.org/`)

- Privacy Engineering by Massachusetts Institute of Technology (MIT) on edX: Learn technical approaches to building privacy-preserving AI systems. (`www.edx.org/school/mitx`)

- Introduction to Artificial Intelligence by Georgia Institute of Technology on Udacity: Gain a foundational understanding of AI concepts for cybersecurity and data privacy applications. (`www.udacity.com/course/intro-to-artificial-intelligence--cs271`)

Market Research and Case Studies

Industry reports and case studies supplement the theoretical knowledge offered in the ebook by providing real-world context. These resources provide insights into how organizations have effectively deployed AI for improved security and privacy, including real examples and lessons gained that can be used to influence strategic decision-making.

Discussion Boards and Community Forums

Community forums and discussion groups provide a venue for professionals, enthusiasts, and experts to exchange thoughts, pose questions, and engage in meaningful debates. These platforms build a feeling of community by allowing readers to draw into collective expertise while also staying up to date on current trends and concerns.

"References and Further Reading" section, in essence, provides a curated roadmap for readers embarking on a rewarding research of AI, information security, and data privacy. Individuals can adapt their learning journey to correspond with their specific interests and professional goals, enabling a constant quest of knowledge in these dynamic and ever-changing domains.

This chapter serves as a springboard for further investigation into AI, information security, and data privacy. It provides a wide range of resources, including foundational texts, research papers, industry reports, and online courses. Readers can delve deeper into specific topics such as ethical considerations, cutting-edge research, and practical applications by using authoritative guidelines, case studies, and community forums. This carefully curated selection enables people to personalize their learning experience and stay current in these ever-changing fields.

Recommended Books, Blogs, etc.

Staying informed and involved in the ever-changing landscape of AI, information security, and data privacy is critical. This section provides a selected selection of recommended resources, including books, blogs, and other platforms, which act as lighthouses for people wishing to enhance their expertise and keep up with the newest advances in the area.

AI and Security Book Recommendations

1. Melanie Mitchell's *Artificial Intelligence: A Guide for Thinking Humans*: This book presents an engrossing investigation of AI's history, current state, and future potential, as well as a thought-provoking viewpoint on the intersections of technology and society.

2. *Data and Goliath: The Hidden Battles to Collect Your Data and Control Your World* by Bruce Schneier: A captivating read for those interested in understanding the broader consequences of data-driven technology, Schneier's work delves into the world of data gathering and monitoring.

3. Kai-Fu Lee's *AI Superpowers: China, Silicon Valley, and the New World Order*: Lee's book investigates the battle between China and Silicon Valley, offering light on the geopolitical and economic sides of AI development.

4. Cennydd Bowles' *Future Ethics*: Bowles examines the ethical considerations of coming technologies, such as AI, and walks readers through a framework for building morally sound digital products and services.

5. Jon Erickson's *Hacking: The Art of Exploitation*: Erickson's book offers a comprehensive reference to hacking techniques, providing hands-on insights into the methods employed by both malevolent actors and security professionals.

Educational Blogs and Online Platforms

1. Schneier on Security (Blog): Written by Bruce Schneier, this blog covers a wide range of security, privacy, and technology themes. It is a useful resource for staying updated because of Schneier's smart commentary and analysis.

2. MIT Technology Review (Online Platform): MIT Technology Review provides in-depth articles and analyses on emerging technologies such as artificial intelligence and cybersecurity. It is a trusted source for the most recent technological breakthroughs.

3. The Privacy Project (*The New York Times*): *The New York Times'* Privacy Project delves into the many facets of privacy in the digital age, with articles, essays, and multimedia content delving into the societal ramifications of data collecting.

4. Dark Reading (Online Platform): Dark Reading is a cybersecurity news, insights, and analysis platform that provides news, insights, and analysis on the newest threats, vulnerabilities, and tactics for securing digital environments.

5. Towards Data Science (Medium Publication): A Medium center for data science and AI aficionados, with a multitude of articles, courses, and case studies ranging from machine learning to data privacy.

Podcasts for In-Depth Debates

1. "Security Now" by TWiT: This podcast, hosted by Steve Gibson and Leo Laporte, covers a wide range of security-related subjects, giving in-depth debates and analysis.

2. "The AI Alignment Podcast" by The Center for the Governance of AI: This podcast investigates the alignment of artificial intelligence with human values, and it features conversations with professionals and thought leaders in the field.

3. "The Privacy, Security, & OSINT Show" by Michael Bazzell: Bazzell's podcast focuses on privacy, security, and open-source intelligence, providing individuals and professionals with practical ideas and insights.

Online Learning Platforms and Courses

1. Andrew Ng's Coursera course "AI for Everyone": This course gives a nontechnical introduction to AI, making it accessible to a broad audience interested in learning the principles of artificial intelligence.

2. Rochester Institute of Technology's edX course "Cybersecurity Fundamentals": This course is ideal for people who want to learn more about cybersecurity. It covers key principles and tactics in the sector.

3. Nishant Bhajaria's "Data Privacy Foundations" course on LinkedIn Learning: Bhajaria's course provides a thorough examination of data privacy principles and best practices.

These recommended resources serve as compass points for anyone wishing to broaden their knowledge, gain practical insights, and contribute to the ongoing dialog shaping these interconnected topics in the dynamic environment of AI, security, and privacy. These materials, whether in the form of books, blogs, podcasts, or online courses, provide a broad and enriching array of perspectives for readers willing to embark on a journey of continual learning.

Glossary: Key Terms and Idea Definitions

Definitions of Key Terms and Concepts

AES: A commonly used symmetric encryption technique that has been established as a standard for secure data encryption.

A sophisticated and long-lasting cyberattack in which an unauthorized person gains access to a network and remains undetected for an extended period of time.

Persistent Attack: A persistent attack is a cyberattack strategy in which an attacker maintains unauthorized access to a computer network for an extended period of time, usually with malicious intent. They move stealthily through the system, stealing data, disrupting operations, or accomplishing other goals while avoiding detection.

Adversarial Robustness: A machine learning model's capacity to withstand and sustain accuracy in the face of intentional input manipulation.

Anomaly Detection: The detection of patterns or behaviors that differ from the norm, which is frequently employed in intrusion detection systems.

Anonymization is the process of eliminating or changing personal information from datasets in order to safeguard individuals' privacy.

Bastion Host: A highly secure server placed on the network's perimeter to protect it from security attacks.

© Ranadeep Reddy Palle, Krishna Chaitanya Rao Kathala 2024
R. R. Palle and K. C. R. Kathala, *Privacy in the Age of Innovation*,
https://doi.org/10.1007/979-8-8688-0461-8

Behavioral Biometrics: The study of patterns of behavior, such as typing speed or mouse movements, in order to authenticate and identify users.

Biometric Encryption: The use of biometric data to encrypt and decrypt data, hence increasing security by connecting cryptographic keys to distinct biological characteristics.

Biometric Template: A mathematical representation of unique biometric traits used in biometric systems for comparison and identification.

Brute Force Attack: A means of gaining unauthorized access by attempting all conceivable password or encryption key combinations.

Cloud security refers to the measures and technologies used to safeguard data, applications, and infrastructure in cloud computing environments.

Consent Management: Obtaining and managing user consent for the collection and processing of personal data in accordance with privacy regulations.

Container security refers to the measures and techniques used to protect containerized applications and their runtime environments, such as Docker or Kubernetes.

Containerization is the process of encapsulating an application and its dependencies in a container to allow for consistent deployment and scalability.

Cross-Origin Resource Sharing (CORS): A security feature that web browsers use to restrict how web pages in one domain request and interact with resources in another domain.

Cross-Site Request Forgery (CSRF): A security flaw in which an attacker baits a user into unintentionally submitting a request to a web application.

Cross-Site Scripting (XSS): A sort of security flaw that allows attackers to inject malicious scripts into web pages that other users are seeing.

Cyber Threat Hunting is the proactive search for indications of malicious activity or security threats within a network or system.

The Dark Web is a section of the Internet that is not indexed by standard search engines and is frequently connected with criminal activity and the trading of unlawful goods and services.

Data Custodian: A human or entity in charge of storing, maintaining, and protecting data inside an organization.

To secure sensitive information, data masking is the technique of disguising original data with fictional or pseudonymous data.

Data Residency: The physical or geographic location of data storage and processing, which is frequently determined by legal and regulatory restrictions.

Deception Technology: The use of bogus assets or information to deceive attackers and identify their presence in a network.

A digital certificate is a cryptographic key pair used to encrypt Internet connections and verify the identity of people or systems.

DDoS: An attack that floods a network, service, or website with traffic in order to overload and interrupt regular operation.

Dynamic Analysis: The study of software behavior in a runtime context in order to uncover and comprehend its functionality as well as potential security risks.

Encryption Algorithm: A set of mathematical principles and techniques for encrypting and decrypting data to ensure secure communication.

Encryption Key Management: The secure generation, storage, distribution, and disposal of encryption keys to prevent unwanted access to sensitive data.

Encryption Key: A cryptographic key that is used to encrypt and decrypt data in order to ensure safe communication and storage.

Endpoint Detection and Response (EDR): A security solution that monitors and responds to advanced endpoint threats, allowing for real-time threat detection.

End-to-End Encryption: A security solution that encrypts data as it travels from its origin to its destination, preventing unauthorized access.

End-User Security Awareness: Individuals are educated and trained to improve their understanding of security threats and best practices.

Ethical hacking refers to hacking actions that are authorized and managed by cybersecurity specialists in order to find vulnerabilities and shortcomings.

Exploit Kit: A set of tools and techniques for automating the exploitation of vulnerabilities in software applications.

A false positive is an improper identification of a security incident or threat that indicates a potential security tool or system problem.

Federated Identity Management: A system that enables users to access many apps or systems across multiple organizations using a single set of credentials.

FIM (Field Integrity Monitoring): The process of evaluating the integrity of files and systems in order to detect unauthorized changes or tampering.

Firewall Rule: A security policy that specifies how a firewall should treat certain types of network traffic.

A firewall is a network security device that monitors and restricts network traffic based on predefined security rules.

Firmware security is the protection of software incorporated in hardware devices against vulnerabilities and illegal alterations.

Firmware: Software embedded in hardware devices that provides low-level control over the hardware on which it operates.

Forensic Analysis: The investigation and analysis of digital data in order to discover and comprehend security events or crimes.

Geofencing is the use of GPS or RFID technology to create a virtual boundary that allows location-based security measures to be implemented.

Ghosting: The unlawful reproduction of a user's identity, which is frequently utilized for harmful purposes.

Governance, Risk, and Compliance (GRC): A comprehensive approach to managing an organization's overall governance, risk management, and regulatory compliance.

Granular Access Control (GAC): Exact control over user permissions, stating which actions or data a user can access.

GPOs (Group Policy Objects): Microsoft Windows operating system settings that define how systems run and interact within a network.

Hash Collision: When two different inputs yield the identical hash value, it might lead to security problems.

Hash Function: A mathematical process that converts input data into a fixed-length string of characters, which is widely used to verify data integrity.

Hashed Password: A password that has been hashed once, making it difficult for attackers to recover the original password.

Hijacking: The unauthorized control or access of a session, connection, or system.

Homomorphic Encryption: An encryption technology that allows computations on encrypted data to be conducted without decrypting it, preserving privacy during processing.

A honeynet is a network of decoy devices that attract and investigate attackers, allowing security professionals to assess their techniques.

A honeypot is a fake system or network that attracts and detects intruders, diverting them away from the genuine target.

A honeytoken is a fake token or piece of data that is placed purposely to detect and notify on unauthorized access or harmful activities.

Incident Command System (ICS): A standardized approach to emergency response command, control, and coordination during security events.

A defined set of processes to follow in the case of a security incident, outlining measures to contain, eradicate, and recover.

ILM (Information Lifecycle Management): Data management that spans the whole data lifecycle, from generation and storage to archiving and disposal.

Controls and technology that limit access and usage of sensitive information even after it has been disseminated are referred to as information rights management (IRM).

Insider Threat: A security threat originating within an organization from individuals who utilize their access to data for nefarious objectives.

Insider trading is defined as the illicit use of sensitive information for financial advantage, which often occurs within a firm.

Intrusion Detection System (IDS): A security tool that monitors network or system activity for malicious behavior or violations of security policies.

Intrusion Prevention System (IPS): A security solution that actively detects and prevents malicious or security threats.

JSON is a lightweight data transfer format that is often used for asynchronous browser/server communication.

JavaScript Security: Preventing security vulnerabilities in JavaScript, a popular programming language in web development.

Jitterbug Attack: A cyberattack that purposefully generates random delays in communication to interrupt a system's normal operation.

Job rotation is a security technique that involves rotating personnel between different tasks or responsibilities on a regular basis to lessen the danger of insider threats.

JSON Injection: An attack that takes advantage of flaws in JSON data handling to perform unauthorized activities on a web application.

JSON Web Token (JWT): A small, URL-safe way of conveying claims between two parties, frequently used for authentication and data transmission.

Jurisdictional Compliance: Adherence to legal standards and laws unique to the geographic place in which an organization works, which influences data handling methods.

Kali Linux: A Linux distribution with a large range of security tools that is specifically designed for penetration testing and ethical hacking.

Kerberos Authentication: A network authentication mechanism that employs tickets to validate user and service identities.

Kerckhoffs' Principle: A cryptographic principle saying that the security of a cryptographic system should be based only on the secrecy of the keys rather than the secrecy of the algorithm.

Key Escrow: The practice of storing encryption keys with a third party to permit access in the event of a legal or emergency demand.

Key Management is the management of cryptographic keys throughout their lives, including generation, distribution, and retirement.

Keylogger: Malicious software or hardware that captures keystrokes and is frequently used to steal sensitive data such as passwords.

Keystroke Dynamics: The study of an individual's unique typing patterns in order to identify and authenticate users based on their typing activity.

Knowledge-Based Authentication (KBA): A technique of verifying an individual's identification that relies on specialized knowledge known only to the individual, such as a password or PIN.

Lagrange Point: A location in space where the gravitational pulls of two big bodies, such as Earth and Moon, form a stable zone for satellite deployment.

Lateral Movement: An attacker's movement across a network in search of unauthorized access to sensitive systems or data.

Least Common Mechanism: A security principle that recommends minimizing shared resources and processes to decrease the effect of potential security breaches.

The principle of granting persons or systems only the minimal degree of access or permissions required to fulfill their tasks.

Logical Access Control (LAC) is the process of restricting access to computer systems based on user credentials, roles, or other logical attributes.

A logical bomb is a piece of code or software that is meant to perform a malicious function when certain conditions are satisfied.

Logical security is the protection of computer systems and data using access controls, encryption, and other software-based measures.

Machine Learning Operations (MLOps): The processes and technology that make it easier to deploy, monitor, and manage machine learning models in production.

Malware analysis refers to the investigation and study of malicious software in order to comprehend its functionality, behavior, and possible impact.

Malware Sandbox: An isolated environment in which malware can be executed and analyzed safely without harming the actual system.

Mandatory Access Control (MAC): A security model in which a central authority determines access permissions based on security labels.

Man-in-the-Middle (MitM) Attack: An attack in which an unauthorized third party intercepts and potentially modifies two parties' communication without their knowledge.

A mantrap is a physical security device that uses a tiny, contained space to check and regulate admission, which is commonly employed in high-security situations.

Metadata: Data descriptive information such as file creation date, authorship, and version history.

Microsegmentation is the process of dividing a network into smaller, isolated segments in order to improve security by managing and monitoring traffic between them.

Network Access Control (NAC) is a security strategy that limits network access based on policies, ensuring that only authorized devices connect.

Network Access Control (NAC) is a security solution that regulates network access based on device compliance with security regulations.

Network Forensics: The investigation and identification of potential risks by analyzing network traffic, logs, and events.

Individual data packets within a network are examined to detect irregularities, intrusions, or security risks.

Network segmentation is the division of a computer network into subnetworks in order to improve security by isolating and reducing the impact of potential security problems.

NTA: The process of scanning and analyzing network traffic in order to detect and respond to security risks.

Non-repudiation: Ensuring that neither the sender nor the recipient can deny sending or receiving a message.

Non-repudiation: The assurance that the sender cannot deny sending a communication and that the recipient cannot dispute receiving it.

OAuth (Open Authorization): A secure authorization protocol that allows third-party applications to access user data without revealing passwords.

OAuth Token: A token that grants a user limited access to a resource. It is often used in authentication and authorization.

Open Source Security: The use of and contribution to open source software in order to improve security, transparency, and collaboration.

OSI Model: A conceptual framework that defines the functions of a telecommunications or computing system.

The Open Web Application Security Project (OWASP) is a nonprofit organization dedicated to improve software security through community-led projects and tools.

Outbound Firewall: A security barrier that monitors and controls incoming and outgoing network traffic in order to prevent illegal data transmission.

Out-of-Band Authentication is the process of verifying a user's identification using a communication channel or mechanism other than the one utilized for the primary transaction.

Packet Sniffing: The unlawful interception and study of data packets transmitted over a network, which may expose sensitive information.

Patch Management is the process of locating, purchasing, testing, and applying fixes to resolve vulnerabilities in software and systems.

Penetration testing is the practice of simulating a cyberattack on a system or network in order to find and solve security flaws.

Physical security refers to safeguards put in place to protect physical assets, facilities, and resources against unlawful access, damage, or theft.

A physical token, such as a smart card or USB token, is a tangible device or object used for authentication.

Privacy by Design: A system design methodology that stresses privacy and data protection throughout the development process.

Privilege Escalation: The process of obtaining greater access or permissions than were originally provided.

Controls and monitors access inside an organization, particularly access granted to users with elevated privileges (PAM).

Quantum communication refers to the application of quantum mechanics concepts for secure communication, such as quantum entanglement and superposition.

Quantum Computer: A type of computing that uses quantum mechanics concepts, potentially affecting cryptography procedures and data security.

Quantum Entanglement: A quantum phenomena in which particles become entangled and the state of one particle influences the state of the other, possibly affecting secure communication.

Quantum Key Distribution (QKD): A secure communication technology that employs quantum physics to enable the secure exchange of encryption keys.

Quantum Key Distribution (QKD): A method of securing communication channels by allowing two parties to generate a shared random secret key using quantum physics.

Quantum Resistant Cryptography: Algorithms that are designed to withstand quantum computer attacks, ensuring the long-term security of encrypted data.

Quarantine (Network Security): Isolating potentially hazardous data or systems to prevent malware or security hazards from spreading.

QR Code: A two-dimensional barcode that may store data and is commonly used for quick access to websites or information.

Blue Team: A cybersecurity exercise in which the red team mimics attackers while the blue team defends against the simulated attacks, thereby improving overall security readiness.

Reverse engineering is the process of disassembling and examining a product in order to learn about its design, architecture, and functionality.

Risk assessment is the assessment of potential hazards and vulnerabilities in order to identify their influence on an organization's goals.

Risk Mitigation: Strategies and measures performed to mitigate or control potential risks and their impact on the operations of a business.

Rogue Device Detection: The detection of unauthorized or unmanaged network devices that may pose security risks.

Root Cause Analysis: The process of identifying and addressing the root cause of a security incident or problem in order to prevent it from happening again.

The basic reason or fault that is accountable for a security incident or system failure.

Rootkit: Malicious software that allows unauthorized access to a computer or network while concealing its presence from users and security software.

SAML stands for Security Assertion Markup Language, and it is an XML-based standard for exchanging authentication and authorization data between parties.

Security Information and Event Management (SIEM): A comprehensive solution that analyzes security alarms generated throughout an organization's IT infrastructure in real time.

SIM (Security Information Management): The collection, analysis, and presentation of security-related data to enable informed decision-making.

SOC (Security Operations Center): A centralized unit in charge of monitoring, detecting, and responding to security incidents.

Security Posture: An organization's total cybersecurity strength and preparedness, indicating its capacity to defend against and respond to threats.

Security Token Service (STS): A service that provides secure authentication and authorization for users and applications by issuing security tokens.

Steganography is the practice of hiding messages or information within non-secret data in order to avoid detection.

Threat hunting is the proactive and iterative search for hidden security threats within a company's network or systems.

Threat intelligence is information about potential or current threats that assists organizations in understanding and mitigating security risks.

Token Ring Network: A LAN topology in which devices are connected in a physical ring or circle.

Tokenization is the practice of replacing sensitive data with a unique identifier or token to prevent unauthorized access to the original material.

Transport Layer Security (TLS): A protocol used in web browsers that ensures privacy between communicating applications and users on the Internet.

Two-Man Rule: A security practice that necessitates the presence of two authorized individuals when performing critical or sensitive tasks, thereby reducing the risk of unauthorized actions.

Unified Threat Management (UTM): A security solution that incorporates numerous security functions into a single integrated platform, such as a firewall, antivirus, and intrusion detection.

Unstructured Data: Information that does not have a predefined data model or organizational structure, such as that found in emails, documents, or social media.

Unsupervised Learning: A machine learning method in which the algorithm is not given labeled training data, allowing it to identify patterns on its own.

User and Entity Activity Analytics (UEBA): The analysis and detection of anomalous patterns in user behavior using machine learning methods.

User Behavior Analytics (UBA): The study of user behavior patterns in order to detect and respond to unusual activities that may indicate a security risk.

User Provisioning: The process of creating, managing, and deactivating user accounts and access rights within the systems of an organization.

Virtual LAN (VLAN): A logical network segmentation used to increase network performance, security, and broadcast efficiency.

Virtual patching is the application of security patches or mitigations to vulnerabilities without modifying the underlying code or software.

A VPN is a secure network connection that allows users to access the internet and share data as if their computing devices were directly connected to a private network.

Vishing: A social engineering attack in which attackers utilize voice conversation to dupe victims into disclosing critical information.

Voice Biometrics: The use of voice patterns to identify and authenticate people, as seen in voice recognition systems.

Vulnerability Disclosure: The responsible reporting and communication of security flaws to software vendors or other parties.

Vulnerability Scanner: Software that finds and evaluates weaknesses in a system or network in order to address security issues proactively.

WAF: A security device or service that filters, monitors, and blocks HTTP traffic between a web application and the Internet.

WAF: A security device or service that filters, monitors, and blocks HTTP traffic in order to protect web applications.

Web Application Security Scanner: A tool that detects security flaws in web applications automatically.

WOT: A system that evaluates and ranks the trustworthiness of websites based on user reviews and ratings.

Worm: A self-replicating piece of malware that spreads across networks and systems, frequently causing damage or draining resources.

X509 Certificate: A standard that defines the format of public key certificates, allowing for secure Internet communication and authentication.

XML Encryption: A standard for encrypting XML data during transmission or storage to ensure confidentiality and integrity.

XML External Entity (XXE) Attack: A type of security flaw in XML parsers that allows an attacker to interfere with XML data processing.

Ain't Markup Language (YAML): A human-readable data serialization format that emphasizes simplicity and readability in configuration files.

YARA Rules: A pattern-matching language used in malware detection and analysis to identify and classify malicious files.

Zero-Knowledge Proof: A cryptographic method that allows one party to prove to another that they are aware of a specific piece of information without revealing it.

Zero-Trust Architecture: A security paradigm that assumes no trust by default, necessitating verification from everyone attempting to access resources, even those within the network.

Zero-Day Exploit: An attack that takes advantage of a software vulnerability on the same day it is discovered by exploiting the lack of available defenses.

Zero-Day Patch: A software vendor's security update that addresses a recently discovered vulnerability before it is exploited.

A zero-day vulnerability is a software flaw that is exploited by attackers on the same day it is made public, frequently before a fix is available.

Index

A

Access controls, 16, 24, 28, 80, 81, 98, 139, 176

Accountability, 88, 113, 134, 144, 149

Adaptive security models, 86

Advanced Persistent Threats (APTs), 26, 29

Adversarial detection systems, 123

Adversarial machine learning, 5, 34, 122
   adaptive defenses, 124
   adversarial attacks, 122
   attackers and defenders, 124
   data security, 124
   defense and countermeasure strategies, 122, 123
   ethical considerations, 123

Adversarial threat mitigation, 96

Adversarial training, 80, 93

AI-driven behavioral analytics, 33, 37

AI-driven compliance
   adaptive compliance strategies, 71
   automated data mapping and classification, 70
   bias and ethical, 74, 75

continuous monitoring and auditing, 71

in data governance processes
   consent management, 72
   data quality management, 72
   incident response and breach detection, 73
   material discovery and classification, 72
   policy enforceability, 72

enhanced collaboration, 77

evolution, 76, 77

improved security with predictive analytics, 70

obstacles and considerations, 75, 76

PIA process
   automated data flow analysis, 73
   collaboration and reporting, 74
   dynamic compliance monitoring, 74
   risk identification and mitigation, 73

risk management and privacy Influence assessments, 71

© Ranadeep Reddy Palle, Krishna Chaitanya Rao Kathala 2024
R. R. Palle and K. C. R. Kathala, *Privacy in the Age of Innovation*,
https://doi.org/10.1007/979-8-8688-0461-8

L

M

Machine learning (ML), 102, 163
    applications, 11
    bias and fairness, 13
    definition, 10
    endpoint protection, 37
    explainability and
        transparency, 13
    federated learning for data
        privacy (*see* Federated
        learning)
    intrusion detection systems, 10
    in malware analysis, 11
    models, 32
    and predictive analytics, 10
    traditional signature-based
        methods, 32
    types, 10
Manufacturing processes, 106, 107
MIT Technology Review, 166
Model deployment security, 96
    anomaly detection, 99
    APIs/endpoints, 100
    authentication/access
        controls, 98
    continuous monitoring, 99
    data encryption, 99
    DevOps security practices, 101
    employee training and
        awareness, 101
    future considerations, 103, 104
    future protection, 104
    importance, 96, 97
    incident response planning, 100
    issues, 97
    model packaging/delivery
        security, 98
    regulatory standards, 100, 101
    software updates/patch
        management, 99
    TLS, 99
    trends and developments,
        102, 103
Model performance
    monitoring, 87
Model poisoning attacks, 59, 81

N

Natural language processing
    (NLP), 34

O

Official guidelines and
    standards, 162
Online platforms and courses, 162,
    163, 167
Orchestration, 35, 121

P

Patch vulnerabilities, 99
Penetration testing, 87, 141
Periodic compliance audits, 84
Personal Data Protection Act
    (PDPA), 67